Biology
Essentials

A Wiley Brand

Biology
Essentials

by Rene Fester Kratz, PhD, and Donna Rae Siegfried with Medhane Cumbay and Traci Cumbay

Biology Essentials For Dummies®

Published by: **John Wiley & Sons, Inc.**, 111 River Street, Hoboken, NJ 07030-5774, www.wiley.com

Copyright © 2019 by John Wiley & Sons, Inc., Hoboken, New Jersey

Published simultaneously in Canada

No part of this publication may be reproduced, stored in a retrieval system or transmitted in any form or by any means, electronic, mechanical, photocopying, recording, scanning or otherwise, except as permitted under Sections 107 or 108 of the 1976 United States Copyright Act, without the prior written permission of the Publisher. Requests to the Publisher for permission should be addressed to the Permissions Department, John Wiley & Sons, Inc., 111 River Street, Hoboken, NJ 07030, (201) 748-6011, fax (201) 748-6008, or online at http://www.wiley.com/go/permissions.

Trademarks: Wiley, For Dummies, the Dummies Man logo, Dummies.com, Making Everything Easier, and related trade dress are trademarks or registered trademarks of John Wiley & Sons, Inc. and may not be used without written permission. John Wiley & Sons, Inc. is not associated with any product or vendor mentioned in this book.

LIMIT OF LIABILITY/DISCLAIMER OF WARRANTY: THE PUBLISHER AND THE AUTHOR MAKE NO REPRESENTATIONS OR WARRANTIES WITH RESPECT TO THE ACCURACY OR COMPLETENESS OF THE CONTENTS OF THIS WORK AND SPECIFICALLY DISCLAIM ALL WARRANTIES, INCLUDING WITHOUT LIMITATION WARRANTIES OF FITNESS FOR A PARTICULAR PURPOSE. NO WARRANTY MAY BE CREATED OR EXTENDED BY SALES OR PROMOTIONAL MATERIALS. THE ADVICE AND STRATEGIES CONTAINED HEREIN MAY NOT BE SUITABLE FOR EVERY SITUATION. THIS WORK IS SOLD WITH THE UNDERSTANDING THAT THE PUBLISHER IS NOT ENGAGED IN RENDERING LEGAL, ACCOUNTING, OR OTHER PROFESSIONAL SERVICES. IF PROFESSIONAL ASSISTANCE IS REQUIRED, THE SERVICES OF A COMPETENT PROFESSIONAL PERSON SHOULD BE SOUGHT. NEITHER THE PUBLISHER NOR THE AUTHOR SHALL BE LIABLE FOR DAMAGES ARISING HEREFROM. THE FACT THAT AN ORGANIZATION OR WEBSITE IS REFERRED TO IN THIS WORK AS A CITATION AND/OR A POTENTIAL SOURCE OF FURTHER INFORMATION DOES NOT MEAN THAT THE AUTHOR OR THE PUBLISHER ENDORSES THE INFORMATION THE ORGANIZATION OR WEBSITE MAY PROVIDE OR RECOMMENDATIONS IT MAY MAKE. FURTHER, READERS SHOULD BE AWARE THAT INTERNET WEBSITES LISTED IN THIS WORK MAY HAVE CHANGED OR DISAPPEARED BETWEEN WHEN THIS WORK WAS WRITTEN AND WHEN IT IS READ.

For general information on our other products and services, please contact our Customer Care Department within the U.S. at 877-762-2974, outside the U.S. at 317-572-3993, or fax 317-572-4002. For technical support, please visit https://hub.wiley.com/community/support/dummies.

Wiley publishes in a variety of print and electronic formats and by print-on-demand. Some material included with standard print versions of this book may not be included in e-books or in print-on-demand. If this book refers to media such as a CD or DVD that is not included in the version you purchased, you may download this material at http://booksupport.wiley.com. For more information about Wiley products, visit www.wiley.com.

Library of Congress Control Number: 2019931684

ISBN: 978-1-119-58958-7 (pbk); ISBN: 978-1-119-58962-4 (ePDF); ISBN: 978-1-119-58954-9 (ePub)

Manufactured in the United States of America

SKY10030962_102721

Contents at a Glance

Table of Contents

Introduction

Life is all around you, from invisible microbes and green plants to the other animals with whom you share the Earth. What's more, these other living things aren't just around you — they're intimately interconnected with your life. Plants make your food and provide you with oxygen, microbes break down dead matter and recycle materials that all living things need, and insects pollinate the plants you rely on for food. Ultimately, all living beings rely on other living beings for their survival.

What makes biology so great is that it allows you to explore the interconnectedness of the world's organisms and really understand that living beings are works of art and machines rolled into one. Organisms can be as delicate as a mountain wildflower or as awe-inspiring as a majestic lion. And regardless of whether they're plants, animals, or microbes, all living things have numerous working parts that contribute to the function of the whole being. They move, obtain energy, use raw materials, and make waste, whether they're as simple as a single-celled organism or as complex as a human being.

Biology is the key you need to unlock the mysteries of life. Through it, you discover that even single-celled organisms have their complexities, from their unique structures to their diverse metabolisms. Biology also helps you realize what a truly miraculous machine your body is, with its many different systems that work together to move materials, support your structure, send signals, defend you from invaders, and obtain the matter and energy you need for growth.

About This Book

Biology Essentials For Dummies takes a look at the characteristics all living things share. It also provides an overview of the concepts and processes that are fundamental to living things. We put an emphasis on looking at how human beings meet their needs, but we also take a look at the diversity of life on planet Earth.

Conventions Used in This Book

To help you find your way through the subjects in this book, we use the following style conventions:

>> *Italics* highlight new words or terms that are defined in the text. They also point out words we want to emphasize.

>> Also, whenever we introduce scientific terms, we try to break the words down for you so that the terms become tied to their meanings, making them easier to remember.

Foolish Assumptions

As we wrote this book, we tried to imagine who you are and what you need in order to understand biology. Here's what we came up with:

>> You're a high school student taking biology, possibly in preparation for an advanced placement test or college entrance examination. If you're having trouble in biology class and your textbook isn't making much sense, try reading the relevant section of this book first to give yourself a foundation and then go back to your textbook or notes.

>> You're a college student who isn't a science major but is taking a biology class to help fulfill your degree requirements. If you want help following along in class, try reading the relevant sections in this book before you go to a lecture on a particular topic. If you need to fix a concept in your brain, read the related section after class.

Icons Used in This Book

We use some of the familiar *For Dummies* icons to help guide you and give you new insights as you read the material. Here's the scoop on what each one means.

REMEMBER

The information highlighted with this icon is stuff we think you should permanently store in your mental biology file. If you want a quick review of biology, scan through the book reading only the paragraphs marked with Remember icons.

TIP

This symbol offers pointers that help you remember the facts presented in a particular section so you can better commit them to memory.

Beyond the Book

In addition to what you're reading right now, this book comes with a free access-anywhere Cheat Sheet. To get this Cheat Sheet, go to www.dummies.com and search for "Biology Essentials For Dummies Cheat Sheet" by using the Search box.

Where to Go from Here

Where you start reading is up to you. However, we do have a few suggestions:

» If you're currently in a biology class and having trouble with a particular topic, jump right to the chapter or section featuring the subject that's confusing you.

» If you're using this book as a companion to a biology class that's just beginning, you can follow along with the topics being discussed in class.

Whatever your situation, the table of contents and index can help you find the information you need.

Chapter **1**

Exploring the Living World

Biology is the study of life, as in the life that covers the surface of the Earth like a living blanket, filling every nook and cranny from dark caves and dry deserts to blue oceans and lush rain forests. Living things interact with all these environments and each other, forming complex, interconnected webs of life.

In this chapter, we give you an overview of the big concepts of biology. Our goal is to show you how biology connects to your life and to give you a preview of the topics we explore in greater detail later in this book.

Living Things: Why Biologists Study Them, What Defines Them

Biologists seek to understand everything they can about living things, including

» The structure and function of all the diverse living things on planet Earth

>> The relationships between living things

>> How living things grow, develop, and reproduce, including how these processes are regulated by DNA, hormones, and nerve signals

>> The connections between living things and their environment

>> How living things change over time

>> How DNA changes, how it's passed from one living thing to another, and how it controls the structure and function of living things

REMEMBER

An individual living thing is called an *organism.* All organisms share eight specific characteristics that define the properties of life:

>> **Living things are made of cells that contain DNA.** A *cell* is the smallest part of a living thing that retains all the properties of life. In other words, it's the smallest unit that's alive. *DNA,* short for *deoxyribonucleic acid,* is the genetic material, or instructions, for the structure and function of cells.

>> **Living things maintain order inside their cells and bodies.** One law of the universe is that everything tends to become random over time. According to this law, if you build a sand castle, it'll crumble back into sand over time. Living things, as long as they remain alive, don't crumble into little bits. They constantly use energy to rebuild and repair themselves so that they stay intact.

>> **Living things regulate their systems.** Living things maintain their internal conditions in a way that supports life. Even when the environment around them changes, organisms attempt to maintain their internal conditions; this process is called *homeostasis.* Think about what happens when you go outside on a cool day without wearing a coat. Your body temperature starts to drop, and your body responds by pulling blood away from your extremities to your core in order to slow the transfer of heat to the air. It may also trigger shivering, which gets you moving and generates more body heat. These responses keep your internal body temperature in the right range for your survival even though the outside temperature is low.

»» Living things respond to signals in the environment. If you pop up suddenly and say "Boo!" to a rock, it doesn't do anything. Pop up and say "Boo!" to a friend or a frog, and you'll likely see him or it jump. That's because living things have systems to sense and respond to signals (or *stimuli*). Many animals sense their environment through their five senses just like you do, but even less familiar organisms, such as plants and bacteria, can sense and respond. For example, during the process of *phototaxis*, plants direct their growth toward areas where they have access to light.

»» Living things transfer energy among themselves and between themselves and their environment. Living things need a constant supply of energy to grow and maintain order. Organisms such as plants capture light energy from the sun and use it to build food molecules that contain chemical energy. Then the plants, and other organisms that eat the plants, transfer the chemical energy from the food into cellular processes. As cellular processes occur, they transfer most of the energy back to the environment as heat.

»» Living things grow and develop. You started life as a single cell. That cell divided to form new cells, which divided again. Now your body is made of approximately 100 trillion cells. As your body grew, your cells received signals that told them to change and become special types of cells: skin cells, heart cells, liver cells, brain cells, and so on. Your body developed along a plan, with a head at one end and a "tail" at the other. The DNA in your cells controlled all these changes as your body developed.

»» Living things reproduce. People make babies, hens make chicks, and plasmodial slime molds make plasmodial slime molds. When organisms reproduce, they pass copies of their DNA onto their offspring, ensuring that the offspring have some of the traits of the parents.

»» Living things have traits that evolved over time. Birds can fly, but most of their closest relatives — the dinosaurs — couldn't. The oldest feathers seen in the fossil record are found on a feathered dinosaur called *Archaeopteryx*. No birds or feathers have been found in any fossils that are older than those of *Archaeopteryx*. From observations like these, scientists can infer that having feathers is a trait that wasn't always present on Earth; rather, it's a trait that developed at a certain point in time. So, today's birds have characteristics that developed through the evolution of their ancestors.

Meet Your Neighbors: Looking at Life on Earth

Life on Earth is incredibly diverse, beautiful, and complex. Heck, you could spend a lifetime exploring the microbial universe alone. The deeper you delve into the living world around you, the more you can appreciate the similarities between all life on Earth — and be fascinated by the differences. The following sections give you a brief introduction to the major categories of life on Earth (called *domains*, as we explain in the upcoming section "Organizing Life into Smaller and Smaller Groups: Taxonomy").

Unsung heroes: Bacteria

Consisting mostly of single-celled organisms, bacteria are *prokaryotic*, meaning they lack a nuclear membrane around their DNA. Most bacteria have a cell wall made of *peptidoglycan:* a hybrid sugar-protein molecule.

REMEMBER

Most people are familiar with disease-causing bacteria such as *Streptococcus pyogenes, Mycobacterium tuberculosis,* and *Staphylococcus aureus.* Yet the vast majority of bacteria on Earth don't cause human diseases. Instead, they play important roles in the environment and health of living things, including humans. Photosynthetic bacteria make significant contributions to planetary food and oxygen production, and *E. coli* living in your intestines make vitamins that you need to stay healthy. So when you get down to it, plants and animals couldn't survive on Earth without bacteria.

Generally speaking, bacteria range in size from 1 to 10 micrometers (one millionth of a meter) in length and are invisible to the naked eye. Along with being nucleus-free, they have a genome that's a single circle of DNA. They reproduce *asexually* (meaning they produce copies of themselves) by a process called *binary fission.*

Bacteria have many ways of getting the energy they need for growth and various strategies for surviving in extreme environments. Their great metabolic diversity has allowed them to colonize just about every environment on Earth.

Bacteria impersonators: Archaeans

Archaeans are prokaryotes, just like bacteria. In fact, you can't tell the difference between the two just by looking, even if you look very closely using an electron microscope, because they're about the same size and shape, have similar cell structures, and divide by binary fission.

Until the 1970s, no one even knew that archaeans existed; up to that point, all prokaryotic cells were assumed to be bacteria. Then, in the 1970s, a scientist named Carl Woese started doing genetic comparisons between prokaryotes. Woese startled the entire scientific world when he revealed that prokaryotes actually separated into two distinct groups — bacteria and archaea — based on sequences in their genetic material.

The first archaeans were discovered in extreme environments (think salt lakes and hot springs), so they have a reputation for being *extremophiles* (*-phile* means "love," so *extremophiles* means "extreme-loving"). Since their initial discovery, however, archaeans have been found everywhere scientists have looked for them. They're happily living in the dirt outside your home right now, and they're abundant in the ocean.

Because archaeans were discovered fairly recently, scientists are still learning about their role on planet Earth, but so far it looks like they're as abundant and successful as bacteria.

A taste of the familiar: Eukaryotes

Unless you're a closet biologist, you're probably most familiar with life in eukaryotic form because you encounter it every day. As soon as you step outside, you can find a wealth of plants and animals (and maybe even a mushroom or two if you look around a little).

On the most fundamental level, all eukaryotes are quite similar. They share a common cell structure with nuclei and organelles, use many of the same metabolic strategies, and reproduce either asexually or sexually.

Despite these similarities, we bet you still feel that you're pretty different from a carrot. You're right to feel that way. The differences between you and a carrot are what separate you into two different kingdoms. In fact, enough differences exist among eukaryotes to separate them into four different kingdoms:

>> **Animalia:** Animals are organisms that begin life as a cell called a *zygote* that results from the fusion of a sperm and an egg. The fertilized egg then divides to form a hollow ball of cells called a *blastula*. If you're wondering when the fur, scales, and claws come into play, these familiar animal characteristics get factored in much later, at the point when animals get divided up into phyla, families, and orders (see the "Organizing Life into Smaller and Smaller Groups: Taxonomy" section later in this chapter for more on these groupings).

>> **Plantae:** Plants are photosynthetic organisms that start life as embryos supported by maternal tissue. This definition of plants includes all the plants you're familiar with: pine trees, flowering plants (including carrots), grasses, ferns, and mosses. All plants have cells with cell walls made of cellulose. They reproduce asexually by mitosis, but they can also reproduce sexually.

The definition of plants, which specifies a stage where an embryo is supported by maternal tissue, excludes most of the algae, like seaweed, found on Earth. Algae and plants are so closely related that many people include algae in the plant kingdom, but many biologists draw the line at including algae in the plant kingdom.

>> **Fungi:** Fungi may look a bit like plants, but they aren't photosynthetic. They get their nutrition by breaking down and digesting dead matter. Their cells have walls made of *chitin* (a strong, nitrogen-containing polysaccharide). This kingdom includes mushrooms, molds that you see on your bread and cheese, and many rusts that attack plants. Yeast is also a member of this kingdom even though it grows differently; most fungi grow as filaments (that look like threads), but yeast grow as little oval cells.

>> **Protista:** This kingdom is defined as everything else that's eukaryotic. Seriously. Biologists have studied animals, plants, and fungi for a long time and defined them as distinct groups long ago. But many, many, eukaryotes don't fit into these three kingdoms. A whole world of microscopic protists exists in a drop of pond water. The protists are so diverse that some biologists think they should be separated into as many as 11 kingdoms of their own.

Classifying Living Things

Much like you'd draw a family tree to show the relationships between your parents, grandparents, and other members of your family, biologists use a *phylogenetic tree* (a drawing that shows the relationships among a group of organisms) to represent the relationships among living things.

Although you probably know how your family members are related to each other, biologists have to use clues to figure out the relationships among living things. The types of clues they use to figure out these relationships include

>> **Physical structures:** The structures that biologists use for comparison may be large, like feathers, or very small, like a cell wall.

>> **Chemical components:** Some organisms produce unique chemicals. Bacteria, for example, are the only cells that make the hybrid sugar-protein molecule called peptidoglycan.

>> **Genetic information:** An organism's genetic code determines its traits, so by reading the genetic code in DNA, biologists can go right to the source of differences among species.

REMEMBER

The more characteristics two organisms have in common with each other, the more closely related they are. Characteristics that organisms have in common are called *shared characteristics*.

REMEMBER

Based on structural, cellular, biochemical, and genetic characteristics, biologists can classify life on Earth into groups that reflect the evolutionary history of the planet. That history indicates that all life on Earth began from one original universal ancestor after the Earth formed 4.5 billion years ago. All the diversity of life that exists today is related because it's descended from that original ancestor.

Organizing Life into Smaller and Smaller Groups: Taxonomy

Biologists need to work with small groups of living things in order to determine how similar the different types of organisms are. Hence, the creation of the *taxonomic hierarchy*, a naming system that ranks organisms by their evolutionary relationships. Within this hierarchy, living things are organized from the largest, most-inclusive group down to the smallest, least-inclusive group.

The taxonomic hierarchy is as follows, from largest to smallest.

>> **Domain:** Domains group organisms by fundamental characteristics such as cell structure and chemistry. For example, organisms in domain Eukarya are separated from those in the Bacteria and Archaea domains based on whether their cells have a nucleus, the types of molecules found in the cell wall and membrane, and how they go about protein synthesis. (We introduce the three domains in the earlier section "Meet Your Neighbors: Looking at Life on Earth.")

>> **Kingdom:** Kingdoms group organisms based on developmental characteristics and nutritional strategy. For example, organisms in the animal kingdom (Animalia) are separated from those in the plant kingdom (Plantae) because of differences in the early development of these organisms and the fact that plants make their own food by photosynthesis, whereas animals ingest their food. (Kingdoms are most useful in domain Eukarya because they're not well defined for the prokaryotic domains.)

>> **Phylum:** Phyla separate organisms based on key characteristics that define the major groups within the kingdom. For example, within kingdom Plantae, flowering plants (Angiosperms) are in a different phylum than cone-bearing plants (Conifers).

>> **Class:** Classes separate organisms based on key characteristics that define the major groups within the phylum. For example, within phylum Angiophyta, plants that have two seed leaves (dicots, class Magnoliopsida) are in a separate class than plants with one seed leaf (monocots, class Liliopsida).

>> **Order:** Orders separate organisms based on key characteristics that define the major groups within the class. For example, within class Magnoliopsida, nutmeg plants (Magnoliales) are put in a different order than black pepper plants (Piperales) due to differences in their flower and pollen structure.

>> **Family:** Families separate organisms based on key characteristics that define the major groups within the order. For example, within order Magnoliales, buttercups (Ranunculaceae) are in a different family than roses (Rosaceae) due to differences in their flower structure.

>> **Genus:** Genera separate organisms based on key characteristics that define the major groups within the family. For example, within family Rosaceae, roses (Rosa) are in a different genus than cherries (Prunus) thanks to differences in their flower structure.

>> **Species:** Species separate eukaryotic organisms based on whether they can successfully reproduce with each other. You can walk through a rose garden and see many different colors of China roses (Rosa chinensis) that are all considered one species because they can reproduce with each other.

TIP

Think of how biologists organize living things like how you might organize your clothing. In your first round of organizing, you might make groups of pants, shirts, socks, and shoes. From there, you might go into the shirt group and organize your shirts into smaller groups, such as short-sleeved versus long-sleeved shirts. Then perhaps you'd organize them by type of fabric, then color, and so on. At some point, you'd have very small groups with very similar articles of clothing — perhaps a group of two short-sleeved, button-down, blue shirts, for example. All your clothing would be organized in a hierarchy, from the big category of clothing all the way down to the small category of short-sleeved, button-down, blue shirts.

Table 1-1 compares the classification, or *taxonomy*, of you, a dog, a carrot, and *E. coli*.

TABLE 1-1 Comparing the Taxonomy of Several Species

Taxonomic Group	Human	Dog	Carrot	E. coli
Domain	Eukarya	Eukarya	Eukarya	Bacteria
Kingdom	Animalia	Animalia	Plantae	Eubacteria
Phylum	Chordata	Chordata	Angiophyta	Proteobacteria
Class	Mammalia	Mammalia	Magnoliopsida	Gammaproteobacteria
Order	Primates	Carnivora	Apiales	Enterobacteriales
Family	Hominidae	Canidae	Apiaceae (Umbelliferae)	Enterobacteriaceae
Genus	*Homo*	*Canus*	*Daucus*	*Escherichia*
Species	*H. sapiens*	*C. familiaris*	*D. carota*	*E. coli*

Of the organisms listed in Table 1-1, you have the most in common with a dog. You're both animals possessing a central nervous chord (phylum Chordata), and you're both mammals (class Mammalia), which means you have hair and the females of your species make milk. However, you also have many differences, including the tooth structure that separates you into the order Primates and a dog into the order Carnivora. If you compare yourself to a plant, you can see that you have certain features of cell structure that place you together in domain Eukarya, but little else in common.

REMEMBER

Two organisms that belong to the same species are the most similar of all. For most eukaryotic organisms, members of the same species can successfully sexually reproduce together, producing live offspring that can also reproduce. Bacteria and archaea don't reproduce sexually, so their species are defined by chemical and genetic similarities.

Biodiversity: Our Differences Make Us Stronger

The diversity of living things on Earth is referred to as *biodiversity*. Almost everywhere biologists have looked on this planet — from the deepest, darkest caves to the lush Amazonian rain forests to the depths of the oceans — they've found life. In the deepest, darkest caves where no light ever enters, bacteria obtain energy from the metals in the rocks. In the Amazonian rain forest, plants grow attached to the tops of trees, collecting water and forming little ponds in the sky that become home to insects and tree frogs. In the deep oceans, blind fish and other animals live on the debris that drifts down to them like snow from the lit world far above. Each of these environments presents a unique set of resources and challenges, and life on Earth is incredibly diverse due to the ways in which organisms have responded to these challenges over time.

The following sections clue you in not only to the reasons why biodiversity is so important and how human actions are harming it, but also how human actions can protect biodiversity moving forward.

Valuing biodiversity

Biodiversity is important — and worth valuing — for the following reasons:

REMEMBER

>> **The health of natural systems depends on biodiversity.** Scientists who study the interconnections between different types of living things and their environments believe that biodiversity is important for maintaining balance in natural systems. Each type of living thing plays a role in its environment, and the loss of even one species can have widespread effects.

>> **Many economies rely upon natural environments.** A whole industry called *ecotourism* has grown up around tour guides leading people on trips through natural habitats and explaining the local biology along the way.

>> **Human medicines come from other living things.** For example, the anticancer drug paclitaxel (Taxol) was originally obtained from the bark of the Pacific yew, and the heart medicine digitalin comes from the foxglove plant.

>> **Biodiversity adds to the beauty of nature.** Natural systems have an aesthetic value that's pleasing to the eye and calming to the mind in today's technologically driven world.

Surveying the threats posed by human actions

As the human population grows and uses more and more of the Earth's resources, the populations of other species are declining as a direct result. Following are the ways in which human actions pose major threats to biodiversity:

>> **Development is reducing the size of natural environments.** People need places to live and farms to raise food. In order to meet these needs, they burn rain forests, drain wetlands, cut down forests, pave over valleys, and plow up grasslands. Whenever people convert land for their own use, they destroy the habitats of other species, causing habitat loss.

>> **Unnatural, human-produced wastes are polluting the air and water.** Automobiles and factories burn gasoline and coal, releasing pollution into the air. Metals from mining and chemicals from factories, farms, and homes get into groundwater. After pollution enters the air and water, it travels around the globe and can hurt multiple species, including humans.

>> **The overharvesting of species to provide food and other materials for human consumption is driving some species to near extinction.** Because they can reproduce, living things such as trees and fish are considered renewable resources. However, if people harvest these resources faster than they can replace themselves, the numbers of individual trees and fish decline. If too few members of a species remain, the survival of that species becomes very unlikely.

>> **Human movements around the globe sometimes carry species into new environments.** An *introduced* (or *non-native*) *species* is a foreign species that's brought into a new environment. Introduced species that are very aggressive and take over habitats are called *invasive species.* Invasive species often have a large environmental impact and cause the numbers of *native species* (organisms belonging to a particular habitat) to decline; they can also attack crop plants and cause human diseases.

Exploring the extinction of species

The combined effects of all the various human actions in Earth's ecosystems are reducing the planet's biodiversity. In fact, the rate of extinctions is increasing along with the size of the human population. No one knows for certain how extensive the loss of species due to human impacts will ultimately be, but there's no question that human practices such as hunting and farming have already caused numerous species to become extinct.

Many scientists believe Earth is experiencing its sixth *mass extinction,* a certain time period in geologic history that shows dramatic losses of many species. (The most famous mass extinction event is the one that occurred about 65 million years ago and included the extinction of the dinosaurs.) Scientists theorize that most of the past mass extinctions were caused by major changes in Earth's climate and that the current extinctions (most recently including black rhinos, Zanzibar leopards, and golden toads) began as a result of human land use but may increase as a result of global warming.

The loss in biodiversity that's currently happening on Earth could have effects beyond just the loss of individual species. Living things are connected to each other and their environment in how they obtain food and other resources necessary for survival. If one species depends on another for food, for example, then the loss of a prey species can cause a decline in the predator species.

The sections that follow introduce you to two classifications of species that biologists are keeping an eye on when it comes to questions of extinction.

Keystone species

REMEMBER

Some species are so connected with other organisms in their environment that their extinction changes the entire composition of species in the area. Species that have such great effects on the balance of other species in their environment are called *keystone species*. As biodiversity decreases, keystone species may die out, causing a ripple effect that leads to the loss of many more species. If biodiversity gets too low, then the future of life itself becomes threatened.

An example of a keystone species is the purple seastar, which lives on the northwest Pacific coast of the United States. Purple seastars prey on mussels in the intertidal zone. When the seastars are present, they keep the mussel population in check, allowing a great diversity of other marine animals to live in the intertidal zone. If the seastars are removed from the intertidal zone, however, the mussels take over, and many species of marine animals disappear from the environment.

Indicator species

REMEMBER

One way biologists can monitor the health of particular environments and the organisms that live in them is by measuring the success of *indicator species:* species whose presence or absence in an environment gives information about that environment.

In the Pacific Northwest region of the United States, the health of old-growth forests is measured by the success of the northern spotted owl, a creature that can make its home and find food only in mature forests that are hundreds of years old. As logging decreases the number and size of these old forests, the number of spotted owls has declined, thereby making the number of spotted owls an indicator of the health, or even the existence, of old-growth forests in the Pacific Northwest. Of course, old-growth forests aren't just home to spotted owls — they shelter a rich diversity of living things, including plants, such as sitka spruce and Western hemlock, and animals, such as elk, bald eagles, and flying squirrels. Old-growth forests also perform important environmental functions such as preventing erosion, floods, and landslides; improving water quality; and providing places for salmon to spawn. If old-growth forests become extinct in the Pacific Northwest, the effects will be far reaching and have many negative impacts on the people and other species in the area.

Protecting biodiversity

Biodiversity increases the chance that at least some living things will survive in the face of large changes in the environment, which is why protecting it is crucial. What can people do to protect biodiversity and the health of the environment in the face of the increasing demands of the human population? No one has all the answers, but here are a few ideas worth trying:

>> Keep wild habitats as large as possible and connect smaller ones with *wildlife corridors* (stretches of land or water that wild animals travel as they migrate or search for food) so organisms that need a big habitat to thrive can move among smaller ones.

>> Use existing technologies and develop new ones to decrease human pollution and clean up damaged habitats. Technologies that have minimal effects on the environment are called *clean* or *green technologies.* Some businesses are trying to use these technologies in order to reduce their impact on the environment.

>> Strive for sustainability in human practices, including manufacturing, fishing, logging, and agriculture. Something that's *sustainable* meets current human needs without decreasing the ability of future generations to meet their needs.

>> Regulate the transport of species around the world so that species aren't introduced into foreign habitats. This step

includes being careful about the transport of not-so-obvious species. For example, ships traveling from one port to another are often asked to empty their ballast water offshore so they don't accidentally release organisms from other waters into their destination harbors.

Making Sense of the World through Observations

The true heart of science isn't a bunch of facts — it's the method that scientists use to gather those facts. Science is about exploring the natural world, making observations using the five senses, and attempting to make sense of those observations. Scientists, including biologists, use two main approaches when trying to make sense of the natural world:

>> **Discovery science:** When scientists seek out and observe living things, they're engaging in *discovery science,* studying the natural world and looking for patterns that lead to new, tentative explanations of how things work (these explanations are called *hypotheses*). If a biologist doesn't want to disturb an organism's habitat, he or she may use observation to find out how a certain animal lives in its natural environment. Making useful scientific observations involves writing detailed notes about the routine of the animal for a long period of time (usually years) to be sure that the observations are accurate.

>> **Hypothesis-based science:** When scientists test their understanding of the world through experimentation, they're engaging in *hypothesis-based science,* which usually calls for following some variation of a process called the scientific method (which we explain in a moment). Modern biologists are using hypothesis-based science to try to understand many things, including the causes and potential cures of human diseases and how DNA controls the structure and function of living things.

Hypothesis–based science can be a bit more complex than discovery science and relies on the scientific method. The *scientific method* is basically a plan that scientists follow while performing scientific experiments and writing up the results. It allows experiments to be duplicated and results to be communicated uniformly. Here's the general process of the scientific method:

1. **Make observations and come up with questions.**

   The scientific method starts when scientists notice something and ask questions like "What's that?" or "How does it work?" just like a child might when he sees something new.

2. **Form a hypothesis.**

   Much like Sherlock Holmes, scientists piece together clues to try to come up with the most likely hypothesis (explanation) for a set of observations. This hypothesis represents scientists' thinking about possible answers to their questions.

   Say, for example, a marine biologist is exploring some rocks along a beach and finds a new worm-shaped creature he has never seen before. His hypothesis is that the creature is some kind of worm.

REMEMBER

   One important point about a scientific hypothesis is that it must be testable, or *falsifiable*. In other words, it has to be an idea that you can support or reject by exploring the situation further using your five senses.

3. **Make predictions and design experiments to test the idea(s).**

   Predictions set up the framework for an experiment to test a hypothesis, and they're typically written as "if . . . then" statements.

   If the marine biologist predicts that the creature he found is a worm, then its internal structures should look like those in other worms he has studied.

4. **Test the idea(s) through experimentation.**

   Scientists must design their experiments carefully in order to test just one idea at a time. As they conduct their experiments, scientists make observations using their five senses and record these observations as their results or data. Scientists conduct multiple tests to ensure that their observations are repeatable.

Continuing with the worm example, the marine biologist tests his hypothesis by dissecting the wormlike creature, examining its internal parts carefully with the assistance of a microscope, and making detailed drawings of its internal structures.

5. Make conclusions about the findings.

Scientists interpret the results of their experiments through *deductive reasoning*, using their specific observations to test their general hypothesis. When making deductive conclusions, scientists consider their original hypothesis and ask whether it could still be true in light of the new information gathered during the experiment. If so, the hypothesis can remain as a possible explanation for how things work. If not, scientists reject the hypothesis and try to come up with an alternate explanation (a new hypothesis) that could explain what they've seen.

In the worm example, the marine biologist discovers that the internal structures of the wormlike creature look very similar to another type of worm he's familiar with. He can therefore conclude that the new animal is likely a relative of that other type of worm.

6. Communicate the conclusions with other scientists.

Communication is a huge part of science. Without it, discoveries can't be passed on, and old conclusions can't be tested with new experiments. When scientists complete some work, they write a paper that explains exactly what they did, what they saw, and what they concluded. Then they submit that paper to a scientific journal in their field. Scientists also present their work to other scientists at meetings, including those sponsored by scientific societies. In addition to sponsoring meetings, these societies support their respective disciplines by printing scientific journals and providing assistance to teachers and students in the field.

Chapter **2**
The Chemistry of Life

Everything that has mass and takes up space, including you and the rest of life on Earth, is made of matter. Atoms make up molecules, which make up the substance of living things. Carbohydrates, proteins, nucleic acids, and lipids are four kinds of molecules that are especially important to the structure and function of organisms. In this chapter, we present a bit of the basic chemistry that's essential for understanding biology.

Exploring Why Matter Matters

Matter is the stuff of life — literally. Every living thing is made of matter. In order to grow, living things must get more matter to build new structures. When living things die, be they plants or animals, microbes such as bacteria and fungi digest the dead matter and recycle it so that other living things can use it again. In fact, pretty much all the matter on Earth has been here since the planet formed 4.5 billion years ago; it has just been recycled since then. So, the stuff that makes up your body may once have been part of *Tyrannosaurus rex*, a butterfly, or even a bacterium.

Following are a few facts you should know about matter:

>> **Matter takes up space.** Space is measured in *volume,* and volume is measured in *liters* (L).

>> **Matter has mass.** *Mass* is the term for describing the amount of matter that a substance has. It's measured in *grams* (g). Earth's gravity pulls on your mass, so the more mass you have, the more you weigh.

>> **Matter can take several forms.** The most familiar forms of matter are solids, liquids, and gases. *Solids* have a definite shape and size, such as a person or a brick. *Liquids* have a definite volume. They can fill a container, but they take the shape of the container that they fill. *Gases* are easy to compress and expand to fill a container.

TIP

To understand the difference between mass and weight, compare your weight on Earth versus your weight on the moon. No matter where you are, your body is made of the same amount of stuff, or matter. But the moon is so much smaller than Earth that it has a lot less gravity to pull on your mass. So, your weight on the moon would be just one-sixth of your weight on Earth, but your mass would remain the same.

The Differences among Atoms, Elements, and Isotopes

All matter is composed of elements. When you break down matter into its smallest components, you're left with individual elements that themselves break down into atoms consisting of even smaller pieces called subatomic particles. And sometimes the number of those subatomic particles within a particular atom differs, creating isotopes. This section has the scoop on all these components of matter.

Tiny, mighty atoms

An *atom* is the smallest whole, stable piece of an element that still has all the properties of that element. It's the smallest "piece" of matter that can be measured.

Here's the basic breakdown of an atom's structure:

REMEMBER

>> **The core of an atom, called the nucleus, contains two kinds of subatomic particles: protons and neutrons.** Both have mass, but only one carries any kind of charge. *Protons* carry a positive charge, but *neutrons* have no charge (they're neutral). Because the protons are positive and the neutrons have no charge, the net charge of an atom's nucleus is positive.

>> **Clouds of electrons surround the nucleus.** *Electrons* carry a negative charge but have almost no mass.

REMEMBER

Atoms become ions when they gain or lose electrons. In other words, *ions* are essentially charged atoms. *Positive (+) ions* have more protons than electrons; *negative (−) ions* have more electrons than protons. Positive and negative charges attract one another, allowing atoms to form bonds, as we explain in the upcoming "Molecules, Compounds, and Bonds" section.

Elements of elements

An *element* is a substance made of atoms that have the same number of protons. Think of them as "pure" substances all made of the same thing.

REMEMBER

Living things use only a handful of the elements in nature. The four most common are hydrogen, carbon, nitrogen, and oxygen, all of which are found in air, plants, and water. (Several other elements exist in smaller amounts in organisms, including sodium, magnesium, phosphorus, sulfur, chlorine, potassium, and calcium.)

I so dig isotopes

All atoms of an element have the same number of protons, but the number of neutrons can change. If the number of neutrons differs between two atoms of the same element, the atoms are called *isotopes* of the element.

For example, carbon-12 and carbon-14 are two isotopes of the element carbon. Atoms of carbon-12 have 6 protons and 6 neutrons. These carbon atoms have a mass number of 12 because their mass is equal to 12. Atoms of carbon-14 still have 6 protons (because all carbon atoms have 6 protons), but they have 8 neutrons, giving them a mass number of 14.

Molecules, Compounds, and Bonds

When you start putting elements together, you get more complex forms of matter, such as molecules and compounds. *Molecules* are made of two or more atoms, and *compounds* are molecules that contain at least two different elements.

TIP

One way to sort out the differences among elements, molecules, and compounds is to think about making chocolate chip cookies. First, you need to mix the wet ingredients: butter, sugar, eggs, and vanilla. Consider each of those ingredients a separate element. You need two sticks of the element butter. When you combine butter plus butter, you get a molecule of butter. Before you add the element of eggs, you need to beat them. So, when you add egg plus egg in a little dish, you get a molecule of eggs. To mix all the wet ingredients together, the molecule of butter is combined with the molecule of eggs, and you get a compound called "wet." Next, you need to mix together the dry ingredients: flour, salt, and baking soda. Think of each ingredient as a separate element. When all the dry ingredients are mixed together, you get a compound called "dry." Only when the wet compound is mixed with the dry compound is the reaction sufficiently ready for the most important element: the chocolate chips.

So what holds the elements of molecules and compounds together? Bonds, of course. Two important types of bonds exist in living things:

>> **Ionic bonds** hold ions joined by their opposite electrical charges. Ionic reactions occur when atoms combine and lose or gain electrons. When sodium (Na) and chlorine (Cl) combine, for example, sodium loses an electron to chlorine. Sodium becomes the positively charged sodium ion (Na^+), and chlorine becomes the negatively charged chloride ion (Cl^-). These two oppositely charged ions are attracted to each other, forming an ionic bond.

>> **Covalent bonds** form when atoms share electrons in a covalent reaction. When two oxygen atoms join to form an oxygen molecule, they share two pairs of electrons with each other. Each shared pair of electrons is one covalent bond, so the two pairs of shared electrons in a molecule of oxygen gas have a double bond. Covalent bonds are extremely important in biology because they hold together the backbones of all biological molecules.

Acids and Bases

Some substances, such as lemon juice and vinegar, have a real edge when you taste them. Others, such as battery acid and ammonia, are so caustic you don't even want to get them on your skin. These substances are acids and bases, both of which have the potential to damage cells.

>> **Acids are molecules that can split apart in water and release hydrogen ions (H^+).** A common example is hydrochloric acid (HCl). When HCl is added to water, it splits apart into H^+ and Cl^-, increasing the number of hydrogen ions in the water/HCl solution.

>> **Bases are molecules that can split apart in water and release hydroxide ions (OH^-).** The most common example is sodium hydroxide (NaOH). When NaOH is added to water, it splits apart into Na^+ and OH^-.

Charged particles, like hydrogen and hydroxide ions, can interfere with the chemical bonds that hold molecules together. Because living things are made of molecules, strong acids and bases can release enough of these ions to cause damage.

The relative concentration of hydrogen to hydroxide ions is represented by the pH scale. The following sections explain the pH scale and how organisms regulate their pH.

"Ph"iguring out the pH scale

The *pH scale* is a system of classifying how acidic or basic a solution is. The term *pH* symbolizes the hydrogen ion concentration in a solution (what proportion of a solution contains hydrogen ions). The pH scale goes from 1 to 14. A pH of 7 is neutral, meaning the amount of hydrogen ions and hydroxide ions in a solution with a pH of 7 is equal, just like in pure water.

A solution that contains more hydrogen ions than hydroxide ions is *acidic*, and the pH of the solution is less than 7. If a molecule releases hydrogen ions in water, it's an acid. The more hydrogen ions it releases, the stronger the acid, and the lower the pH value. A solution that contains more hydroxide ions than hydrogen ions is *basic*, and its pH is higher than 7.

Buffing up on buffers

In organisms, blood and cytoplasm are the "solutions" in which the required ions (for example, electrolytes) float. That's why most substances in the body hover around the neutral pH of 7. However, nothing's perfect, so the human body has a backup system in case things go awry. A system of buffers exists to help neutralize the blood if excess hydrogen or hydroxide ions are produced.

REMEMBER

Buffers keep solutions at a steady pH by combining with excess hydrogen (H^+) or hydroxide (OH^-) ions. Think of them as sponges for hydrogen and hydroxide ions. If a substance releases these ions into a buffered solution, the buffers will "soak up" the extra ions.

The most common buffers in the human body are bicarbonate ion (HCO_3^-) and carbonic acid (H_2CO_3). Bicarbonate ion carries carbon dioxide through the bloodstream to the lungs to be exhaled, but it also acts as a buffer. Bicarbonate ion takes up extra hydrogen ions, forming carbonic acid and preventing the pH of the blood from going too low. If the opposite situation occurs and the pH of the blood gets too high, carbonic acid breaks apart to release some hydrogen ions, which brings the pH back into balance.

Carbon-Based Molecules: The Basis for All Life

All living things rely pretty heavily on one particular type of molecule: carbon. The little ol' carbon atom, with its six protons and an outer shell of four electrons, is the central focus of *organic chemistry*, which is the chemistry of living things. When carbon bonds to hydrogen (which happens frequently in organic molecules), the carbon and hydrogen atoms share a pair of electrons in a covalent bond. Molecules with a lot of carbon–hydrogen bonds are called *hydrocarbons*. Nitrogen, sulfur, and oxygen are also often joined to carbon in organisms.

So where do the carbon-containing molecules come from? The answer's simple: food. Some living things, like people, need to eat other living things to get their food, but some organisms, like plants, can make their own food. Regardless of the food source, all living things use food as a supply of carbon-containing molecules.

Carbon atoms are central to all organisms because they're found in carbohydrates, proteins, nucleic acids, and lipids — otherwise known as the structural materials of all living things. The sections that follow describe the roles of these materials.

Providing energy: Carbohydrates

Carbohydrates, as the name implies, consist of carbon, hydrogen, and oxygen. The basic formula for carbohydrates is CH_2O, meaning the core structure of a carbohydrate is one carbon atom, two hydrogen atoms, and one oxygen atom. This formula can be multiplied; for example, glucose has the formula $C_6H_{12}O_6$, which is six times the ratio but still the same basic formula.

REMEMBER

Carbohydrates are energy-packed compounds. Living creatures can break carbohydrates down quickly, making them a source of near-immediate energy. However, the energy supplied by carbohydrates doesn't last long. Therefore, reserves of carbohydrates in the body must be replenished frequently, which is why you find yourself hungry every four hours or so. Although carbohydrates are a source of energy, they also serve as structural elements (such as cell walls in plants).

Carbohydrates come in the following forms:

>> **Monosaccharides:** Simple sugars consisting of three to seven carbon atoms are *monosaccharides* (see Figure 2-1a). In living things, monosaccharides form ring-shaped structures and can join together to form longer sugars. The most common monosaccharide is glucose.

>> **Disaccharides:** Two monosaccharide molecules joined together form a *disaccharide* (see Figure 2-1b). Common disaccharides include sucrose (table sugar) and lactose (the sugar found in milk).

>> **Oligosaccharides:** More than two and up to ten monosaccharides joined together are an *oligosaccharide* (see Figure 2-1c). Oligosaccharides are important markers on the outsides of your cells, such as the oligosaccharides that determine whether your blood type is A or B. (People with type O blood don't have any of this particular oligosaccharide.)

>> **Polysaccharides:** Long chains of monosaccharide molecules linked together form a *polysaccharide* (see Figure 2-1d). Some of these babies are huge, and when we say huge, we mean some of them can have thousands of monosaccharide molecules joined together. Starch and glycogen, which serve as a means of storing carbohydrates in plants and animals, respectively, are examples of polysaccharides.

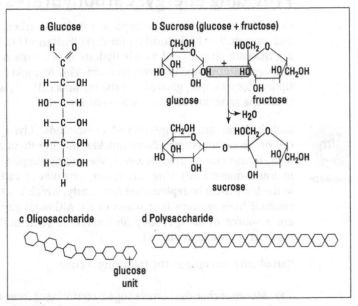

FIGURE 2-1: A variety of carbohydrate molecules.

The next sections explain how sugars interact with one another and how the human body stores a particular carbohydrate known as glucose.

Making and breaking sugars

Monosaccharides join together in a process known as *dehydration synthesis*, which involves two molecules bonding and losing a water molecule. Figure 2-1b shows the dehydration synthesis of glucose and fructose to form sucrose.

The term *dehydration synthesis* may sound technical, but it's not at all if you really think about what the words mean. *Dehydration* is what happens when you don't drink enough water. You dry out because water is removed (but not completely) from some cells,

TIP

such as those in your tongue, to make sure more important cells, like those in your heart or brain, continue to function. *Synthesis* means making something. In dehydration synthesis, something must be made when water is removed. When glucose and fructose get together, a water molecule is removed from the monosaccharides and given off as a byproduct of the reaction.

The opposite of dehydration synthesis is hydrolysis. A *hydrolysis* reaction breaks down a larger sugar molecule into its original monosaccharides. When something undergoes hydrolysis, a water molecule splits a compound (*hydro* means "water"; *lysis* means "break apart"). When sucrose is added to water, it splits apart into glucose and fructose.

Converting glucose for storage purposes

Carbohydrates are found in nearly every food, not just bread and pasta. Fruits, vegetables, and even meats also contain carbohydrates, although meats don't contain very many. Basically, any food that contains sugar has carbohydrates, and most foods are converted to sugars when they're digested.

When you digest your food, the carbohydrates from it break down into small sugars such as glucose. Those glucose molecules are then absorbed from your intestinal cells into your bloodstream, which carries the glucose molecules throughout your entire body. The glucose enters each of your body's cells and is used as a source of carbon and energy.

Because glucose provides a rapid source of energy, organisms often keep some on hand. They store it in various polysaccharides that can be quickly broken down when glucose is needed. Consider the following list your primer on the things glucose can be stored as:

>> **Glycogen:** Animals, including people, store a polysaccharide of glucose called *glycogen*. It has a compact structure, so cells can store a lot of it for later use. Your liver, in particular, keeps a large glycogen reserve on hand for when you exercise.

>> **Starch:** Plants store glucose as the polysaccharide *starch*. The leaves of a plant produce sugar during photosynthesis and then store some of that sugar as starch. When the simple sugars need to be retrieved for use, the starch is broken down into its smaller components.

Plants also make a polysaccharide of glucose called *cellulose.* Cellulose plays a structural role for plants rather than a storage role by giving rigidity to the walls of plant cells. Most animals, including people, can't digest cellulose because of the type of bonds between the glucose molecules. Because cellulose passes through your digestive tract virtually untouched, it helps maintain the health of your intestines.

Making life possible: Proteins

Without proteins, living things wouldn't exist. Many proteins provide structure to cells; others bind to and carry important molecules throughout the body. Some proteins are involved in reactions in the body when they serve as enzymes. Still others are involved in muscle contraction or immune responses. Proteins are so diverse that we can't possibly tell you about all of them. What we can tell you about, however, are the basics of their structure and their most important functions.

The building blocks of proteins

Amino acids, of which there are 20, are the foundation of all proteins. Think of them as train cars that make up a train called a protein. Figure 2-2 shows an amino acid.

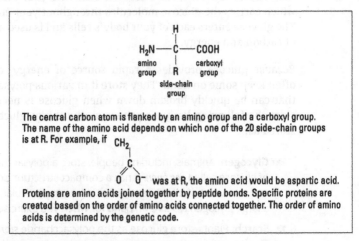

The central carbon atom is flanked by an amino group and a carboxyl group. The name of the amino acid depends on which one of the 20 side-chain groups is at R. For example, if CH_2 ... was at R, the amino acid would be aspartic acid. Proteins are amino acids joined together by peptide bonds. Specific proteins are created based on the order of amino acids connected together. The order of amino acids is determined by the genetic code.

FIGURE 2-2: Amino acid structure.

The genetic information in cells (the DNA) calls for amino acids to link together in a certain order, forming chains called *polypeptide chains.* Amino acids link together by dehydration synthesis, just like sugars do (as we explain in the earlier "Making and breaking sugars" section), and each polypeptide chain is made up of a unique number and order of amino acids.

The main functions of proteins

One or more polypeptide chains come together to form functional *proteins.* Once formed, each protein does a specific job or makes up a specific tissue in the body:

>> **Enzymes are proteins that speed up the rate of chemical reactions.** Metabolic processes, such as the breakdown of carbohydrates or the production of proteins, don't happen automatically; they require enzymes.

>> **Structural proteins reinforce cells and tissues.** *Collagen,* a structural protein found in connective tissue, is the most abundant protein in animals with a backbone. Connective tissue includes ligaments, tendons, cartilage, bone tissue, and even the cornea of the eye. It provides support in the body, and it has a great capability to be flexible and resistant to stretching.

>> **Transport proteins move materials around cells and around the body.** *Hemoglobin* is a transport protein in red blood cells that carries oxygen around the body. A hemoglobin molecule is shaped kind of like a three-dimensional four-leaf clover without a stem. Each leaf of the clover is a separate polypeptide chain. In the center of the clover, but touching each polypeptide chain, is a heme group with an atom of iron at its center. When gas exchange occurs between the lungs and a blood cell, the iron atom attaches to the oxygen. Then, the iron-oxygen complex releases from the hemoglobin molecule in the red blood cell so the oxygen can cross cell membranes and get inside any cell of the body.

Drawing the cellular road map: Nucleic acids

Nucleic acids are large molecules that carry tons of small details, specifically all the genetic information for an organism. Nucleic acids exist in every living thing — plants, animals, bacteria, and fungi. Just think about that fact for a moment. People may look different than fungi, and plants may behave differently than bacteria, but deep down all living things contain the same chemical "ingredients" making up very similar genetic material.

REMEMBER

Nucleic acids are made up of strands of *nucleotides*. Each nucleotide has three components:

>> A nitrogen-containing base called a *nitrogenous base*

>> A sugar that contains five-carbon molecules

>> A phosphate group

That's it. Your entire genetic composition, personality, and maybe even your intelligence hinge on molecules containing a nitrogen compound, some sugar, and a phosphate. The following sections introduce you to the two types of nucleic acids.

Deoxyribonucleic acid (DNA)

You may have heard DNA (short for *deoxyribonucleic acid*) referred to as "the double helix." That's because DNA contains two strands of nucleotides arranged in a way that makes it look like a twisted ladder. See for yourself in Figure 2-3.

The sides of the ladder are made up of sugar and phosphate molecules, hence the nickname "sugar-phosphate backbone." (The name of the sugar in DNA is deoxyribose.) The "rungs" on the ladder of DNA are made from pairs of nitrogenous bases from the two strands.

REMEMBER

The nitrogenous bases that DNA builds its double helix upon are adenine (A), guanine (G), cytosine (C), and thymine (T). The order of these chemical letters spells out your genetic code. Oddly enough, the bases always pair in a certain way: Adenine always goes with thymine (A–T), and guanine always links up with cytosine (G–C). These particular *base pairs* line up just right chemically so that hydrogen bonds can form between them.

Certain sections of nitrogenous bases along a strand of DNA form a gene. A *gene* is a unit that contains the genetic information or codes for a particular protein and transmits hereditary information to the next generation. Whenever a new cell is made in an organism, the genetic material is reproduced and put into the new cell. The new cell can then create proteins and also pass on the genetic information to the next new cell.

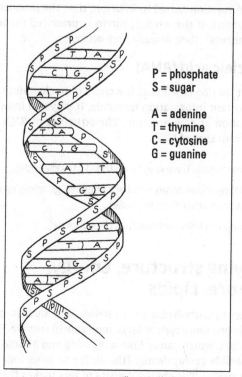

P = phosphate
S = sugar

A = adenine
T = thymine
C = cytosine
G = guanine

Illustration by Kathryn Born, MA

FIGURE 2-3: The twisted-ladder model of a DNA double helix.

But genes aren't only in reproductive cells. Every cell in an organism contains DNA (and therefore genes) because every cell needs to make proteins. Proteins control function and provide structure. Therefore, the blueprints of life are stored in each and every cell.

The order of the nitrogenous bases on a strand of DNA (or in a section of the DNA that makes up a gene) determines the order in which amino acids are strung together to make a protein. Which protein is produced determines which structural element is produced within your body (such as muscle tissue, skin, or hair) or what function can be performed (such as the transportation of oxygen to all the cells).

REMEMBER

Every cellular process and every aspect of metabolism is based on genetic information stored in DNA and thus the production of the proper proteins. If the wrong protein is produced (as in the case of some cancers), then disease may occur.

Ribonucleic acid (RNA)

RNA, short for *ribonucleic acid*, is a chain of nucleotides that serves as an important information molecule. It plays an important role in the creation of new proteins. The structure of RNA is slightly different from that of DNA.

>> RNA molecules have only one strand of nucleotides.

>> The nitrogenous bases used are adenine, guanine, cytosine, and uracil (rather than thymine).

>> The sugar in RNA is ribose (not deoxyribose).

Supplying structure, energy, and more: Lipids

In addition to carbohydrates, proteins, and nucleic acids, your body needs one more type of large molecule to survive. We're talking about fats, which can be both a blessing and a curse because of their incredible *energy density* (the ability to store lots of calories in a small space). The energy density of fats makes them a highly efficient way for living things to store energy — very useful when food isn't always available. But that same energy density makes it really easy to pack in the calories when you eat fatty foods!

Fats are an example of a type of molecule called lipids. *Lipids* are *hydrophobic* molecules, meaning they don't mix well with water.

REMEMBER

Three major types of lipid molecules exist:

>> **Phospholipids:** These lipids, made up of two fatty acids and a phosphate group, have an important structural function for living things because they're part of the membranes of

cells. Phospholipids aren't the type of lipid floating around the bloodstream clogging arteries.

>> **Steroids:** These lipid compounds, consisting of four connecting carbon rings and a functional group that determines the steroid, generally create hormones. *Cholesterol* is a steroid molecule used to make testosterone and estrogen; it's also found in the membranes of cells. The downside to cholesterol is that it's transported around the body by other lipids. If you have too much cholesterol floating in your bloodstream, then you have an excess of fats carrying it. This situation is troubling because the fats and cholesterol molecules can get stuck in your blood vessels, leading to blockages that cause heart attacks or strokes.

>> **Triglycerides:** These fats and oils, which are made up of three fatty acid molecules and a glycerol molecule, are important for energy storage and insulation. In people, fats form from an excess of glucose. After the liver stores all the glucose it can as glycogen, whatever remains is turned into triglycerides. (Both sugars and fats are made of carbon, hydrogen, and oxygen, so your cells just rearrange the atoms to convert from one to another.) The triglycerides float through your bloodstream on their way to be deposited into *adipose tissue* — the soft, squishy fat you can see on your body. Adipose tissue is made up of many, many molecules of fat. The more fat molecules that are added to the adipose tissue, the bigger the adipose tissue (and the place on your body that contains it) gets.

Whether a triglyceride is a fat or an oil depends on the bonds between the carbon and hydrogen atoms.

- Fats contain lots of single bonds between their carbon atoms. These *saturated bonds* pack tightly (see Figure 2-4), so fats are solid at room temperature.

- Oils contain lots of double bonds between their carbon atoms. These *unsaturated bonds* don't pack tightly (see Figure 2-4), so oils are liquid at room temperature.

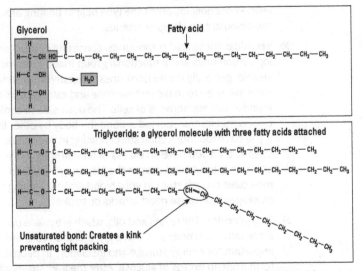

FIGURE 2-4: Saturated and unsaturated bonds in a typical triglyceride.

Fat provides an energy reserve to your body. When you use up all of your stored glucose (which doesn't take long because sugars "burn" quickly in aerobic conditions), your body starts breaking down glycogen, which is stored primarily in the liver and muscle. Liver glycogen stores can typically last 12 or more hours. After that, your body starts breaking down adipose tissue to retrieve some stored energy. That's why aerobic exercise, so long as it's enough to use up more calories than you took in that day, is the best way to lose fat.

Chapter **3**
The Living Cell

Every living thing has cells. The smallest creatures have only one, yet they're as alive as you are. In plain and simple terms, a cell is the smallest living piece of an organism — including you. Without cells, you'd be a disorganized blob of chemicals that'd ooze out into the environment.

You get to explore the purpose and structure of cells in this chapter. And because cells rely on chemical reactions to make things happen, you also find out all about *enzymes,* which are proteins that help speed up the pace of chemical reactions.

An Overview of Cells

Cells are sacs of fluid that are reinforced by proteins and surrounded by membranes. Inside the fluid float chemicals and *organelles,* structures that are used during metabolic processes such as the production or breakdown of proteins.

REMEMBER

A cell is the smallest part of an organism that retains characteristics of the entire organism. For example, a cell can take in fuel, convert it to energy, and eliminate wastes, just like the organism as a whole can. Because cells can perform all the functions of life (as Figure 3-1 shows), the cell is the smallest unit of life.

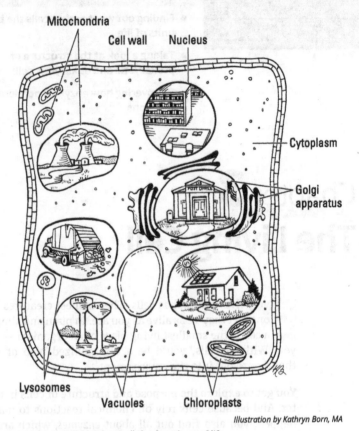

FIGURE 3-1: Cells perform all the functions of life.

Illustration by Kathryn Born, MA

Cells can be categorized in different ways, according to structure or function, or in terms of their evolutionary relationships. In terms of structure, scientists categorize cells based on their internal organization:

>> **Prokaryotes** don't have a "true" nucleus in their cells, nor do they have organelles. Bacteria and archaea are all prokaryotes.

>> **Eukaryotes** have a nucleus in their cells that houses their genetic material. They also have organelles. Plants, animals, algae, and fungi are all eukaryotes.

Peeking at Prokaryotic Cells

Prokaryotes include cells you've probably heard of, such as the bacteria *E. coli* and *Streptococcus* (which causes strep throat), the blue-green algae that occasionally cause lake closures, and the live cultures of bacteria in yogurt, as well as some cells you may never have heard of, called archaeans. Whether you've heard of a specific prokaryote or not, you're likely well aware that bacteria have a pretty bad rap. They seem to make the papers only when they're causing problems, such as disease. Behind the scenes, though, bacteria are quietly performing many beneficial tasks for life on planet Earth. Why, if bacteria could get some good headlines, those headlines might read a little something like this:

>> **Bacteria are used in human food production!** Yogurt and cheese are quite tasty, humans say.

>> **Bacteria can clean up our messes!** Oil-eating bacteria help save beaches; other bacteria help clean up our sewage.

>> **Normal body bacteria help prevent disease!** Bacteria living on the body can prevent disease-causing bacteria from moving in.

>> **Bacteria are nature's recyclers!** Bacteria release nutrients from dead matter during decomposition.

>> **Bacteria help plants grow!** Nitrogen-fixing bacteria can pull nitrogen out of the air and convert it to a form that plants can use.

The cells of prokaryotes are fairly simple in terms of structure because they don't have internal membranes or organelles like eukaryotic cells do. (We cover all the structures present in eukaryotic cells later in this chapter.) Most prokaryotic cells (like the one in Figure 3-2) share these characteristics:

>> A plasma membrane forms a barrier around the cell, and a rigid cell wall outside the plasma membrane provides additional support to the cell.

>> DNA, the genetic material of prokaryotes, is located in the cytoplasm, in an area called the nucleoid.

>> Ribosomes make proteins in the cytoplasm.

>> Prokaryotes break down food using cellular respiration and another type of metabolism called *fermentation* (which doesn't require oxygen).

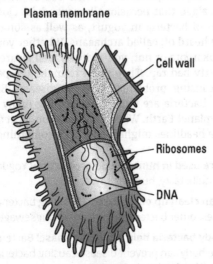

Illustration by Kathryn Born, MA

FIGURE 3-2: A prokaryotic cell.

Examining the Structure of Eukaryotic Cells

The living things you're probably most familiar with — humans, animals, plants, mushrooms, and molds — are all eukaryotes, but they're not the only members of the eukaryote family. Eukaryotes also include many inhabitants of the microbial world, such as algae, amoebas, and plankton.

Eukaryotes have the following characteristics (see Figures 3-3 and 3-4 for diagrams of eukaryotic cells):

>> A nucleus that stores their genetic information or DNA.

>> A plasma membrane that encloses the cell and separates it from its environment.

>> Internal membranes, such as the endoplasmic reticulum and the Golgi apparatus, which create specialized compartments inside the cells.

>> A cytoskeleton made of proteins that reinforces the cells and controls cellular movements.

>> Organelles called *mitochondria* that combine oxygen and food to transfer the energy from food to a form that cells can use.

>> Organelles called *chloroplasts,* which use energy from sunlight plus water and carbon dioxide to make food. (Chloroplasts are found only in the cells of plants and algae.)

>> A rigid cell wall outside of their plasma membrane. (This is found only in the cells of plants, algae, and fungi; animal cells just have a plasma membrane, which is soft.)

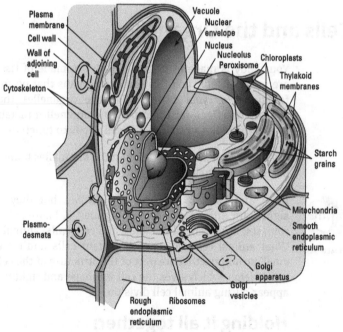

Plasma membrane
Cell wall
Wall of adjoining cell
Cytoskeleton
Vacuole
Nuclear envelope
Nucleus
Nucleolus
Peroxisome
Chloroplasts
Thylakoid membranes
Starch grains
Mitochondria
Smooth endoplasmic reticulum
Golgi apparatus
Golgi vesicles
Plasmo-desmata
Rough endoplasmic reticulum
Ribosomes

FIGURE 3-3: Structures in a typical plant cell.

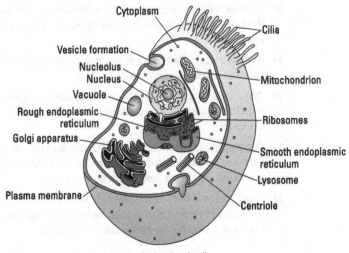

FIGURE 3-4: Structures in a typical animal cell.

Cells and the Organelles

Your body is made of organs, which are made of tissues, which are made of cells. Just like you have organs that perform specific functions for your body, cells have organelles that perform specific functions for the cell. Some organelles metabolize food; others make the structures the cell needs to function.

The sections that follow highlight the organelles found in eukaryotic cells and their specific functions.

REMEMBER

Plant and animal cells are very similar, but they have a few significant differences in their organelles. Plant cells have chloroplasts, large central vacuoles, and cell walls; animal cells don't. What animal cells *do* have that plant cells don't are *centrioles*, small structures that are part of the structure of the cell called the *cytoskeleton*, which gives the cell its shape and rigidity. Centrioles appear during animal cell division.

Holding it all together: The plasma membrane

The membrane that encloses and defines all cells as separate from their environment is called the *plasma membrane*, or the *cell membrane.* The job of the plasma membrane is to separate the chemical reactions occurring inside the cell from the chemicals outside the cell.

TIP

Thinking of the plasma membrane as an international border controlling what enters and leaves a particular country is a good way of remembering the plasma membrane's function.

The fluid inside a cell, called the *cytoplasm,* contains all the organelles and is very different from the fluid found outside the cell. (*Cyto* means "cell," and *plasm* means "shape." So, *cytoplasm* literally means "cell shape," which is fitting because the plasma membrane is what defines cell shape.)

Animal cells are supported by a fluid protein-and-carbohydrate matrix called the *extracellular matrix.* (*Extra* means "outside," so *extracellular* literally means "outside the cell.") Plant cells are supported by a more solid structure, called a *cell wall,* that's made of the carbohydrate cellulose.

The next sections explain the structure of the plasma membrane in detail and describe how materials move through it in order to keep the cell healthy and allow it to do its job.

Deciphering the fluid-mosaic model

Plasma membranes are made of several different components, much like a mosaic work of art. Because membranes are a mosaic, and because they're flexible and fluid, scientists call the description of membrane structure the *fluid-mosaic model.* We've drawn the model for you in Figure 3-5 to help you visualize all the parts that make up a plasma membrane.

Notice the phospholipid bilayer highlighted on the left side of Figure 3-5. This serves as the foundation of the plasma membrane. *Phospholipids* are a special kind of lipid; they have water-attracting *and* water-repelling parts. At body temperature, phospholipids have the consistency of thick vegetable oil, which allows plasma membranes to be flexible and fluid. Each phospholipid molecule has a hydrophilic head that's attracted to water and a hydrophobic tail that repels water. (*Hydro* means "water," *phile* means "love," and *phobia* means "fear," so *hydrophilic* literally means "water-loving" and *hydrophobic* literally means "water-fearing.")

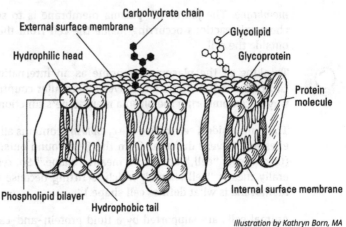

Carbohydrate chain
External surface membrane
Glycolipid
Hydrophilic head
Glycoprotein
Protein molecule
Phospholipid bilayer
Internal surface membrane
Hydrophobic tail

Illustration by Kathryn Born, MA

FIGURE 3-5: The fluid-mosaic model of plasma membranes.

REMEMBER

In each cell, the hydrophilic heads point toward the watery environments outside and inside the cell, sandwiching the hydrophobic tails between them to form the phospholipid bilayer (see Figure 3-5). Because cells reside in a watery solution (the extracellular matrix), and because they contain a watery solution inside of them (cytoplasm), the plasma membrane forms a sphere around each cell so that the water-attracting heads are in contact with the fluid, and the water-repelling tails are protected on the inside.

In addition to phospholipids, proteins are a major component of plasma membranes. The proteins are embedded in the phospholipid bilayer, but they can drift in the membrane like ships sailing through an oily ocean.

Cholesterol and carbohydrates are minor components of plasma membranes, but they play fairly significant roles.

» Cholesterol makes the membrane more stable and prevents it from solidifying when your body temperature is low. (It keeps you from literally freezing when you're "freezing.")

» Carbohydrate chains attach to the outer surface of the plasma membrane on each cell. When carbohydrates attach to the phospholipids, they form glycolipids (and when they attach to the proteins, they form glycoproteins). Your DNA determines which specific carbohydrates attach to your cells, affecting characteristics such as your blood type.

Transporting materials through the plasma membrane

Cells are busy places. They manufacture materials that need to be shipped out, and they take up materials such as food. These important exchanges take place at the plasma membrane.

Whether a molecule can cross a plasma membrane depends upon its structure and the cell. Small, hydrophobic molecules such as oxygen and carbon dioxide are compatible with the hydrophobic tails of the phospholipid bilayer, so they can easily scoot across membranes. Hydrophilic molecules such as ions can't get through the tails by themselves, so they need help to cross. Larger molecules (think food and hormones) also need help, which comes in the form of transport proteins.

Some of these proteins help form openings called *channels* in the membrane. Small molecules such as hormones and ions may be allowed to pass through these *protein channels*. Other proteins, called *carrier proteins*, pick up molecules (the sugar glucose, for example) on one side of a membrane and then drop them off on the other side. Other proteins in the membrane act as *receptors* that detect the presence of different types of molecules, such as the hormone insulin. When specific molecules bind to their receptors, they produce a response in the cell. For example, the binding of a specific molecule with its receptor on muscle cells causes muscle contraction.

REMEMBER

Because the plasma membrane is choosy about what substances can pass through it, it's said to be selectively permeable. (Permeability describes the ease with which substances can pass through a border, such as a cell membrane. *Permeable* means that most substances can easily pass through. *Impermeable* means substances can't pass through. *Selectively permeable* and *semipermeable* mean that only certain substances can pass through.)

Materials can pass through the plasma membrane either passively or actively.

PASSIVELY MOVING ALONG

Passive transport requires no energy on the part of the cell. Molecules move passively across membranes in one of two ways. In both cases, the molecules move from where they're more concentrated to where they're less concentrated. (In other words, they spread themselves out randomly until they're evenly distributed.) Here are the two methods of passive transport:

>> **Diffusion:** The movement of molecules other than water from an area where they're highly concentrated to an area where they're less concentrated is *diffusion.*

>> **Osmosis:** The movement of water across a membrane is *osmosis.* It works the same way as diffusion, but it can be a little confusing because the movement of water is affected by the concentration of substances called *solutes* that are dissolved in the water. Basically, water moves from areas where it's more concentrated (more pure) to areas where it's less concentrated (where it has more solutes).

ACTIVELY HELPING MOLECULES ACROSS

Active transport requires some energy from the cell to move molecules that can't cross the phospholipid bilayer on their own from where they're less concentrated to where they're more concentrated. Carrier proteins, called *active transport proteins* or protein *pumps,* use energy stored in the cell to concentrate molecules inside or outside of the cell.

TIP

Active transport is a little like having to pay to take the Staten Island Ferry. The ferry is the carrier protein, and you're the big molecule that needs help getting from the bloodstream (New York Bay) to the inside of the cell (New York City). The fee that you pay is equivalent to the energy molecules expended by the cell.

Supporting the cell: The cytoskeleton

Much like your skeleton reinforces the structure of your body, the *cytoskeleton* of a cell reinforces that cell's structure. However, it provides that reinforcement in the form of protein cables rather than bones. The proteins of the cytoskeleton reinforce the plasma membrane and the nuclear envelope (covered in the next section). They also run through the cell like railroad tracks, helping organelles circulate around the cell.

TIP

Think of the cytoskeleton as a cell's scaffolding and railroad tracks because it reinforces the cell and allows things to move around within it.

REMEMBER

Some cells have whiplike projections that help them swim or move fluids. If the projections are short, like those shown in Figure 3-4, the structures are called *cilia*. If the projections are long, they're called *flagella*. Both cilia and flagella contain cytoskeletal proteins. The proteins flex back and forth, making the cilia and flagella beat like little whips. Cells with cilia exist in your respiratory tract, where they wiggle their cilia to move mucus so you can cough it out; they're also found in your digestive tract, where they help move food along. Flagella are present on human sperm cells; they enable sperm to swim rapidly toward an egg during sexual reproduction.

The nucleus controls the show

Every cell of every living thing contains genetic material called DNA. In eukaryotic cells, DNA is contained within a chamber called a *nucleus* that's separated from the cytoplasm by a membrane called the *nuclear envelope* (also known as the *nuclear membrane*). In the nucleus of cells that aren't multiplying, the DNA is wound around proteins and loosely spread out in the nucleus. When DNA is in this form, it's called *chromatin*. However, right before a cell divides, the chromatin coils up tightly into chromosomes. Human cells have 46 *chromosomes*, each one of which is a separate piece of DNA. You got 23 chromosomes from your mom and 23 from your dad for a total of 46.

REMEMBER

DNA contains the instructions for building molecules, mostly proteins, that do the work of the cell. Cell function depends upon the action of these proteins, and organism function depends on cell function. So, ultimately, organism function depends upon the instructions in the DNA.

TIP

Consider the nucleus the library of the cell because it holds lots of information. The chromosomes are the library's books, full of instructions for building cells.

Proteins in the nucleus copy the instructions from the DNA into molecules of RNA that get shipped out to the cytoplasm, where they direct the behavior of the cell. For example, each nucleus has a round mass inside it called a *nucleolus*. The nucleolus produces ribosomes, which move out to the cytoplasm to help make proteins. In experiments where scientists transplant the nucleus from one cell into the cytoplasm of another cell, the cell behaves according to the instructions in the nucleus. So, the nucleus is the true control center of the cell.

Creating proteins: Ribosomes

Ribosomes are small structures in the cytoplasm of cells. The instructions for proteins are copied from the DNA into a new molecule called *messenger RNA* (mRNA). The mRNA leaves the nucleus and carries the instructions to the ribosomes in the cytoplasm of the cell. The ribosomes then organize the mRNA and other molecules necessary to build proteins.

Thinking of ribosomes like workbenches where proteins are built is a good way to remember their function.

TIP

Serving as the cell's factory: The endoplasmic reticulum

The *endoplasmic reticulum* (ER) is a series of canals that connects the nucleus to the cytoplasm of the cell. (*Endo* means "inside," and *reticulum* refers to the netlike appearance of the ER, so *endoplasmic reticulum* basically means "a netlike shape inside the cytoplasm.") As you can see in Figure 3-4, part of the ER is covered in dots, which are actually ribosomes that attach to it during the synthesis of certain proteins. This part is called the *rough ER*, or RER. The part of the ER without ribosomes is called the *smooth ER* (SER).

Ribosomes on the RER make proteins that either get shipped out of the cell or become part of the membrane. (Proteins that stay in the cell are put together on ribosomes that float free in the cytoplasm.) The SER is involved in the metabolism of *lipids* (fats). Proteins and lipids made at the ER get packaged up into little spheres of membrane called *transport vesicles* that carry the molecules from the ER to the nearby Golgi apparatus (which we explain in a moment).

To remember the ER's purpose, think of the ER as the cell's internal manufacturing plant because it produces proteins and lipids and then ships them out (to the Golgi apparatus).

TIP

Preparing products for distribution: The Golgi apparatus

The *Golgi apparatus*, which is located very close to the ER (as you can see in Figure 3-4), looks like a maze with water droplets splashing off of it. The "water droplets" are transport vesicles bringing material from the ER to the Golgi apparatus.

Inside the Golgi apparatus, products produced by the cell, such as hormones or enzymes, are chemically tagged and packaged for export either to other organelles or to the outside of the cell. After the Golgi apparatus has processed the molecules, it packages them back up into a vesicle and ships them out again. If the molecules are to be shipped out of the cell, the vesicle finds its way to the plasma membrane, where certain proteins allow a channel to be produced so that the products inside the vesicle can be secreted to the outside of the cell. Once outside the cell, the materials can enter the bloodstream and be transported through the body to where they're needed.

Consider the Golgi apparatus the cell's post office because it receives molecular packages and tags them for shipping to their proper destination.

Cleaning up the trash: Lysosomes

Lysosomes are special vesicles formed by the Golgi apparatus to clean up the cell. Lysosomes contain digestive enzymes, which are used to break down products that may be harmful to the cell and "spit" them back out into the extracellular fluid. (We fill you in on enzymes in the later "Presenting Enzymes, the Jump-Starters" section.) Lysosomes also remove dead organelles by surrounding them, breaking down their proteins, and releasing them to construct a new organelle.

Essentially, lysosomes are the waste collectors of the cell; they gather materials the cell no longer needs and break them down so some parts can be recycled.

Destroying toxins: Peroxisomes

Peroxisomes are little sacs of enzymes that break down many different types of molecules and help protect the cell from toxic products. Peroxisomes help in the breakdown of lipids, making their energy available to the cell.

Some of the reactions that occur in peroxisomes produce hydrogen peroxide, which is a dangerous molecule to cells. Peroxisomes prevent your cells from being damaged by hydrogen peroxide by converting the hydrogen peroxide into plain old water plus an extra oxygen molecule, both of which are always needed by the body and can be used in any cell.

TIP

Peroxisomes are a little bit like food processors. They're involved in breaking things down, just like the blades of a food processor are used to chop up larger pieces of food.

Providing energy, ATP-style: Mitochondria

Mitochondria supply cells with the energy they need to move and grow by breaking down food molecules, extracting their energy, and transferring it to an energy-storing molecule that cells can easily use. That energy-storing molecule is *ATP*, short for *adenosine triphosphate*.

TIP

Think of mitochondria as the power plants of the cell because they produce the energy the cell needs.

REMEMBER

The process mitochondria use to transfer the energy in foods to ATP is called *cellular respiration*. What occurs during cellular respiration is like what occurs when a campfire burns, just on a much smaller scale. In a campfire, wood burns, consuming oxygen and transferring energy (heat and light) and matter (carbon dioxide and water) to the environment. In a mitochondrion, food molecules break down, consuming oxygen and transferring energy to cells (to be stored in ATP) and the environment (as heat).

Converting energy: Chloroplasts

Chloroplasts are organelles found almost exclusively in plants and algae. They specialize in transferring energy from the sun into the chemical energy in food. They often have a distinctly green color because they contain *chlorophyll*, a green pigment that can absorb sunlight. During photosynthesis, the energy from sunlight is used to combine the atoms from carbon dioxide and water to produce sugars like glucose, from which all types of food molecules can be made.

TIP

Consider chloroplasts the plant equivalent of solar-powered kitchens because they use energy from the sun and "ingredients" from the environment (carbon dioxide and water) to make food.

REMEMBER

A very common misconception is that plants have chloroplasts rather than mitochondria. The truth is, plants have both! Think about it — it wouldn't do plants much good to make food if they couldn't also break it down. When plants make food, they store matter and energy for later. When they need the matter and energy, they use their mitochondria to break the food down into usable energy.

Presenting Enzymes, the Jump-Starters

Chemical reactions occur whenever the molecules in cells change. They're usually part of a cycle or pathway that has separate reactions at each step. Of course, because the pace of life in cells is so fast, cells can't just wait around for chemical reactions to happen — they have to make them happen quickly. Fortunately, they have the perfect tool at their disposal in the form of proteins called enzymes.

REMEMBER

Each reaction of a pathway or cycle requires a specific enzyme to act as a *catalyst*, something that speeds up the rate of chemical reactions. These proteins are folded in just the right way to do a specific job. Enzymes also have pockets, called *active sites*, which they use to attach to certain molecules. The molecule an enzyme binds to is called its *substrate* (see Figure 3-6).

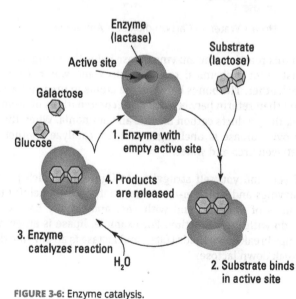

FIGURE 3-6: Enzyme catalysis.

Without the specific enzyme necessary to catalyze a particular reaction, the cycle or pathway can't be completed. The result of an uncompleted pathway is the lack of what that pathway is supposed to produce (a *product*). Without a needed product, a function can't be performed, which negatively affects the organism. For example, if people don't get enough vitamin C, the enzymes needed to make collagen can't function, resulting in a disease called scurvy. The lack of collagen in people with scurvy causes bleeding gums, loss of teeth, and abnormal bone development in children.

The sections that follow introduce you to how enzymes work, what they need to get the job done, and how cells are able to keep them under control.

Staying the same . . .

REMEMBER

Enzymes themselves are recycled. They're the same at the end of a reaction as they were at the beginning, and they can do their job again. For example, the first enzymatic reaction discovered was the one that breaks down urea into products that can be excreted from the body. The enzyme urease catalyzes the reaction between the reactants urea and water, yielding the products carbon dioxide and ammonia, which can be excreted easily by the body.

Urease

Urea + Water ↔ Carbon dioxide + Ammonia

In this reaction, the enzyme urease helps the *reactants* (molecules that enter a chemical reaction), urea and water, combine with each other. The bonds between the atoms in urea and water break and then reform between different combinations of atoms, forming the products carbon dioxide and ammonia. When the reaction is over, urease is unchanged and can catalyze another reaction between urea and water.

TIP

If you find yourself struggling to figure out which proteins are enzymes and which enzymes do what, here's a helpful hint: The names of enzymes end with -*ase* and usually have something to do with their function. For example, lipase is an enzyme that helps break down lipids (fats), and lactase is an enzyme that helps break down lactose.

. . . while lowering activation energy

Enzymes work by reducing the amount of *activation energy* needed to start a reaction so reactions can occur more easily. On their own, reactants could occasionally collide with each other the right way to start a reaction. But they wouldn't do it nearly often enough to keep up with the fast pace of life in a cell. Without enzymes, your body wouldn't be able to, say, clear urea out of your body fast enough, leading to a toxic buildup of urea. That's where the enzyme urease comes into play. It binds the reactants in its active site and brings them together in a way that requires less energy for them to react.

Because reactions can occur more easily with enzymes, they occur more often. This increases the overall rate of the reaction in the body. One way to understand how enzymes speed up reactions is to think about reactions in terms of energy. In order for a reaction to occur, the reactants must collide with enough energy to get the reaction going. In the urea and water example, the reactants would need to collide with each other in just the right way for them to exchange partners and form into carbon dioxide and ammonia.

REMEMBER

Don't fall for the idea that enzymes add energy to reactions to make them happen. They don't. In fact, they don't add *anything* to a reaction; they just help the reactants get together in the right way, lowering the "barrier" to the reaction. In other words, enzymes don't add energy; they just make it so the reactants have enough energy on their own.

Getting some help from cofactors and coenzymes

Enzymes are proteins, but many need a nonprotein partner in order to do their job. Inorganic partners, such as iron, potassium, magnesium, and zinc ions, are called *cofactors*. Organic partners are called *coenzymes*; they're small molecules that can separate from the protein component of the enzyme and participate directly in the chemical reaction. Examples of coenzymes include many derivatives of vitamins. An important function of coenzymes is that they transfer electrons, atoms, or molecules from one enzyme to another.

Controlling enzymes through feedback inhibition

Cells manage their activity by controlling their enzymes via *feedback inhibition*, a process in which a reaction pathway proceeds normally until the final product is produced at too high of a level. The final product then binds to the allosteric site of one of the initial enzymes in the pathway, shutting it down. (An *allosteric site* is literally an "other shape" site. When molecules bind to these "other" pockets, enzymes can be shut down.) By controlling enzymes, cells regulate their chemical reactions and, ultimately, the physiology of the entire organism.

TIP

Feedback inhibition gets its name because it uses a feedback loop. The quantity of the final product provides feedback to the beginning of the pathway. If the cell has plenty of the final product, then the cell can stop running the pathway.

By inhibiting the activity of an initial enzyme, the entire pathway is stopped. The process of feedback inhibition prevents cells not only from having to use energy creating excess products but also from having to make room to store the excess products. It's like keeping yourself from spending money on a huge quantity of food that you won't eat and would just end up storing until it rots.

REMEMBER

Feedback inhibition is reversible because the binding of the final product to the enzyme isn't permanent. In fact, the final product is constantly binding, letting go, and then binding again. When the cell uses up its stores of the final product, the enzyme's allosteric site is empty, and the enzyme becomes active again.

Chapter **4**
Energy and Organisms

Just like you need to put gas in your car's engine so your car can move, you need to put food in your body so it can function. And you're not alone. Every person, as well as every other living thing, needs to "fill its tank" with matter and energy in the form of food. Food molecules are used to build the molecules that make up cells and are broken down to release energy to cells so they can grow and maintain themselves. Animals obtain their food by eating plants and other animals, whereas plants make their own food. In this chapter, we present some facts about the various types of energy and how they're transferred. We also demonstrate why cells need energy and take a look at how cells obtain and then store energy and matter.

What's Energy Got to Do with It?

Whether you realize it or not, you use energy every day to cook your food, brighten your home, and run your appliances. And, really, that's what energy is — something that allows work to be done.

You can probably think of many kinds of energy in your life: electricity, heat, light, and chemicals (like gasoline). Although they may seem very different, the kinds of energy you can think of represent the two main types of energy:

>> **Potential energy** is the energy that's stored in something because of the way it's arranged or structured. Energy in a battery, water behind a dam, and a stretched rubber band about to be released are all examples of potential energy. Food and gasoline also contain potential energy called *chemical potential energy* (energy that's stored in the bonds of molecules).

>> **Kinetic energy** is the energy of motion. Light, heat, and moving objects all contain kinetic energy.

The following sections get you acquainted with the rules surrounding energy. They also explain how the cells of living things use and transfer energy, as well as how they obtain it. (Here's a hint: It's all about food.)

Looking at the rules regarding energy

Energy has three specific rules that are helpful to know so you can better understand how organisms use it:

>> **Energy can't be created or destroyed.** The electricity that people get from hydroelectric power (or coal-burning power plants, wind turbines, or solar panels) isn't created from nothing. It's actually transferred from some other kind of energy. And when people use, say, electricity, that energy doesn't disappear. Instead, it becomes other kinds of energy, such as light or heat.

REMEMBER

The idea that energy can't be created or destroyed is known as the *First Law of Thermodynamics.*

>> **Energy is transferred when it moves from one place to another.** To understand this rule, picture a flowing river that's used as a source of hydroelectric power. Energy from the moving river is transferred first to a spinning turbine, then to flowing electrons in power lines, and finally to the lights shining in customers' homes.

>> **Energy is transformed when it changes from one form to another.** Again, think about a hydroelectric power plant. The potential energy of the water behind the plant's dam is

transformed first into the kinetic energy of moving water, then the kinetic energy of a spinning turbine, and finally the kinetic energy of moving electrons.

Metabolizing molecules

Organisms follow the rules of physics and chemistry, and the human body is no exception. The First Law of Thermodynamics (explained in the preceding section) applies to your *metabolism*, which is all the chemical reactions occurring in your cells at one time.

Two types of chemical reactions can occur as an organism metabolizes molecules:

>> **Anabolic reactions** build molecules. Specifically, small molecules are combined into large molecules for repair, growth, or storage. Like the building of proteins (large molecule) from amino acids (small molecules).

>> **Catabolic reactions** break down molecules like sugars, fats, or proteins to release their stored energy.

REMEMBER

During chemical reactions, atoms receive new bonding partners, and energy may be transferred. (For more on molecules, atoms, and chemical bonds, flip to Chapter 2.)

REMEMBER

Each type of food molecule you're familiar with — carbohydrates, proteins, and fats — is a large molecule that can be broken down into smaller subunits. Complex carbohydrates, also called *polysaccharides*, break down into simple sugars called *monosaccharides*; proteins break down into *amino acids*; and fats and oils break down into *glycerol* and *fatty acids*. After cells break down large food molecules into their subunits, they can more easily reconnect the subunits into the specific molecules that they need.

Transferring energy with ATP

Cells transfer energy between anabolic and catabolic reactions by using an energy middleman — *adenosine triphosphate* (or ATP for short). Energy from catabolic reactions is transferred to ATP, which then provides energy for anabolic reactions.

ATP has three phosphates attached to it (*tri-* means "three," so *triphosphate* means "three phosphates"). When ATP supplies energy to a process, one of its phosphates gets transferred to another molecule, turning ATP into *adenosine diphosphate* (ADP). Cells re-create ATP by using energy from catabolic reactions to reattach a phosphate group to ADP. Cells constantly build and break down ATP, creating the ATP/ADP cycle shown in Figure 4-1.

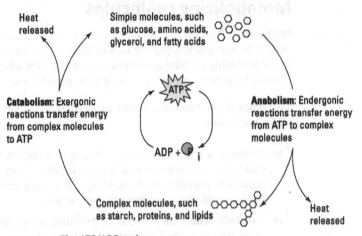

Heat released

Simple molecules, such as glucose, amino acids, glycerol, and fatty acids

Catabolism: Exergonic reactions transfer energy from complex molecules to ATP

ATP

Anabolism: Endergonic reactions transfer energy from ATP to complex molecules

ADP + P_i

Complex molecules, such as starch, proteins, and lipids

Heat released

FIGURE 4-1: The ATP/ADP cycle.

Cells have large molecules that contain stored energy, but when they're busy doing work, they need a handy source of energy. That's where ATP comes in. Cells keep ATP on hand to supply energy for all the work that they do.

TIP

Think of ATP like cash in your pocket. You may have money deposited in the bank, but that money isn't always easy to get, which is why you keep some cash in your pocket to quickly buy what you need. After you spend all your cash, you have to go back to the bank or an ATM to get more. For living things, the energy stored in large molecules is like money in the bank. Cells break down ATP just like you spend your cash. Then, when cells need more ATP, they have to go back to the bank of large molecules and break some more down.

Consuming food for matter and energy

Food molecules — in the form of proteins, carbohydrates, and fats — provide the matter and energy that every living thing needs to fuel anabolic and catabolic reactions and create ATP. (For more about matter and molecules, see Chapter 2.)

>> **Organisms need matter to build their cells so they can grow, repair themselves, and reproduce.** Imagine that you scrape your knee and actually remove a fair amount of skin. Your body repairs the damage by building new skin cells to cover the scraped area. Just like a person who builds a house needs wood or bricks, your body needs molecules to build new cells. (Head to Chapter 3 for the full scoop on cells.)

>> **Organisms need energy so they can move, build new materials, and transport materials around their cells.** These activities are all examples of *cellular work*, the energy-requiring processes that occur in cells. When you walk up stairs, the muscle cells in your legs contract, and each contraction uses some energy. But the activities you decide to engage in aren't the only things that require energy. Your individual cells need energy to do their work.

REMEMBER

Food is a handy package that contains two things every organism needs: matter and energy.

Finding food versus producing your own

All organisms need food, but there's one major difference in how they approach this problem: Some organisms, such as plants, can make their own food; other organisms, like you, have to eat other organisms to obtain their food. Biologists have come up with two separate categories to highlight this difference in how living things obtain their food:

>> **Autotrophs (also known as *producers*) can make their own food.** *Auto* means "self," and *troph* means "feed," so *autotrophs* are self-feeders. Plants, algae, and green bacteria are all examples of autotrophs.

>> **Heterotrophs (also known as *consumers*) have to eat other organisms to get their food.** *Hetero* means "other," so *heterotrophs* are quite literally other-feeders. Animals, fungi, and most bacteria are examples of heterotrophs.

Although you may think that obtaining food is as easy as heading to the supermarket, pulling up to a drive-through window, or meeting the delivery guy at the front door, acquiring nutrients is actually a metabolic process. More specifically, food is made through one process and broken down through another. These processes are as follows:

>> **Photosynthesis:** Only autotrophs such as plants, algae, and green bacteria engage in *photosynthesis,* a process that consists of using energy from the sun, carbon dioxide from the air, and water from the soil to make sugars. (The carbon dioxide provides the matter plants need for food building.) When plants remove hydrogen atoms from water to use in the sugars, they release oxygen as waste.

>> **Cellular respiration:** Both autotrophs and heterotrophs do *cellular respiration,* a process that uses oxygen to help break down food molecules such as sugars. The energy stored in the bonds of the food molecules is transferred to ATP. As the energy is transferred to the cells, the matter from the food molecules is released as carbon dioxide and water.

REMEMBER

If you think about it, photosynthesis and cellular respiration are really the opposites of each other. Photosynthesis consumes carbon dioxide and water, producing food and oxygen. Cellular respiration consumes food and oxygen, producing carbon dioxide and water. Scientists write the big picture view of both processes as the following equations:

Photosynthesis:

$$6 CO_2 + 6 H_2O + \text{Light Energy} \rightarrow C_6H_{12}O_6 + 6 O_2$$

Cellular respiration:

$$C_6H_{12}O_6 + 6 O_2 \rightarrow 6 CO_2 + 6 H_2O + \text{Usable Energy (ATP)}$$

REMEMBER

Don't fall for the idea that only heterotrophs such as animals engage in cellular respiration. Autotrophs such as plants do it too. Think of it like this: Photosynthesis is a food-making pathway that autotrophs use to store matter and energy for later. So, a plant doing photosynthesis is like you packing a lunch for yourself. There wouldn't be much point in packing the lunch if you weren't going to eat it later, right? The same is true for a plant. It does photosynthesis to store matter and energy. When it needs that matter and energy, it uses cellular respiration to "unpack" its food.

Building Cells through Photosynthesis

Autotrophs such as plants combine matter and energy to make food in the form of sugars. With those sugars, plus some nitrogen and minerals from the soil, autotrophs can make all the types of molecules they need to build their cells. The chemical formula for *glucose*, the most common type of sugar found in cells, is $C_6H_{12}O_6$. To build glucose, autotrophs need carbon, hydrogen, and oxygen atoms, plus energy to combine them into sugar.

>> The carbon and oxygen for the sugars come from carbon dioxide in Earth's atmosphere.

>> The hydrogen for the sugars comes from water in the soil.

>> The energy to build the sugars comes from the sun (but only in autotrophs that use photosynthesis).

A common misconception is that plants get the matter they need to grow from the soil. The reality is that plants get most of the matter they need to grow from the carbon dioxide in the air. This idea may be more difficult to believe because air, including carbon dioxide, doesn't seem like much of anything, but scientists have proven that it's correct. Plants collect a lot of carbon dioxide molecules (CO_2) from the air and combine them with water molecules (H_2O) to build sugars such as glucose ($C_6H_{12}O_6$). Plants get the water they need, plus some small amounts of minerals such as nitrogen, from the soil.

Photosynthesis occurs in two main steps (Figure 4-2 depicts both in action):

REMEMBER

>> **The light reactions of photosynthesis transform light energy into chemical energy.** The chemical energy is stored in the energy carrier ATP and in NADPH: a molecule that can store energy in the form of electrons and is used in the production of lipids and nucleic acids.

>> **The light-independent reactions of photosynthesis produce food.** ATP from the light reactions supplies the energy needed to combine carbon dioxide (CO_2) and water (H_2O) to make glucose ($C_6H_{12}O_6$).

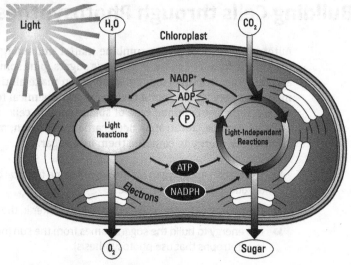

Light | H_2O | Chloroplast | CO_2

NADP+
ADP
+ P

Light Reactions

Light-Independent Reactions

ATP
NADPH

Electrons

O_2 | Sugar

FIGURE 4-2: The two halves of photosynthesis, the light reactions and the light-independent reactions, are separate but linked.

The next sections delve deeper into the process of photosynthesis.

Transforming energy from the ultimate energy source

The sun is a perfect energy source: a nuclear reactor positioned at a safe distance from planet Earth. It contains all the energy you could ever need . . . if only you could capture it. Well, green bacteria figured out how to do just that more than 2.5 billion years ago, showing that photosynthetic autotrophs were way ahead of humans on this one.

Plants, algae, and green bacteria use pigments to absorb light energy from the sun. You're probably most familiar with the pigment *chlorophyll*, which colors the leaves of plants green. The chloroplasts in plant cells contain lots of chlorophyll in their membranes so they can absorb light energy (see Chapter 3 for more on chloroplasts).

During the light reactions of photosynthesis, chloroplasts absorb light energy from the sun and then transform it into the chemical energy stored in ATP. When the light energy is absorbed, it splits water molecules. The electrons from the water molecules help with the energy transformation from light energy to chemical energy in ATP. Plants release the oxygen from the water molecules as waste, producing the oxygen (O_2) that you breathe.

Putting matter and energy together

Plants use the energy in ATP (which is a product of the light reactions) to combine carbon dioxide molecules and water molecules to create glucose during the light-independent reactions. To make glucose, plants first take carbon dioxide out of the air through a process called *carbon fixation* (taking carbon dioxide and attaching it to a molecule inside the cell). They then use the energy from the ATP and the electrons that came from water to convert the carbon dioxide to sugar.

REMEMBER

As their name indicates, the light-independent reactions of photosynthesis don't need direct sunlight to occur. However, plants need the products of the light reactions to run the light-independent reactions, so really, the light-independent reactions can't happen if the light reactions can't happen.

When plants have made more glucose than they need, they store their excess matter and energy by combining glucose molecules into larger carbohydrate molecules, such as starch. When necessary, plants can break down the starch molecules to retrieve glucose for energy or to create other compounds, such as proteins and nucleic acids (with added nitrogen taken from the soil) or fats (many plants, such as olives, corn, peanuts, and avocados, store matter and energy in oils).

Cellular Respiration: Using Oxygen to Break Down Food

Autotrophs and heterotrophs do cellular respiration to break down food to transfer the energy from food to ATP. The cells of animals, plants, and many bacteria use oxygen to help with the energy transfer during cellular respiration; in these cells, the type of cellular respiration that occurs is aerobic respiration (*aerobic* means "with air").

REMEMBER

Three separate pathways combine to form the process of cellular respiration (you can see them all in action in Figure 4-3). The first two, glycolysis and the Krebs cycle, break down food molecules. The third pathway, oxidative phosphorylation, transfers the energy from the food molecules to ATP. Here are the basics of how cellular respiration works:

>> During *glycolysis,* which occurs in the cytoplasm of the cell, cells break glucose down into *pyruvate,* a three-carbon compound. After glycolysis, pyruvate is broken down into a two-carbon molecule called acetyl-coA.

>> After pyruvate is converted to acetyl-coA, cells use the *Krebs cycle* (which occurs in the matrix of the mitochondrion) to break down acetyl-coA into carbon dioxide.

>> During *oxidative phosphorylation,* which occurs in the inner membrane or *cristae* of the mitochondrion, cells transfer energy from the breakdown of food to ATP.

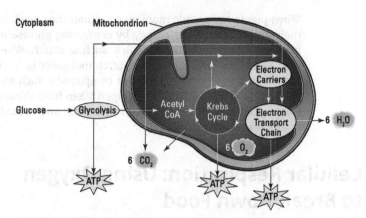

FIGURE 4-3: An overview of cellular respiration.

For a more in-depth look at cellular respiration, check out the following sections.

REMEMBER

Cellular respiration is different from plain ol' respiration. *Respiration,* which is more commonly referred to as breathing, is the physical act of inhaling and exhaling. *Cellular respiration* is what happens inside cells when they use oxygen to transfer energy from food to ATP.

Breaking down food

After the large molecules in food are broken down into their smaller subunits, the small molecules can be further broken down to transfer their energy to ATP. During cellular respiration, enzymes slowly rearrange the atoms in food molecules. Each rearrangement produces a new molecule in the pathway and can also produce other useful molecules for the cell. Some reactions

>> **Release energy that can be transferred to ATP:** Cells quickly use this ATP for cellular work, such as building new molecules.

>> **Oxidize food molecules and transfer electrons and energy to coenzymes:** *Oxidation* is the process that removes electrons from molecules; *reduction* is the process that gives electrons to molecules. During cellular respiration, enzymes remove electrons from food molecules and then transfer the electrons to the coenzymes nicotinamide adenine dinucleotide (NAD^+) and flavin adenine dinucleotide (FAD). NAD^+ and FAD receive the electrons as part of hydrogen (H) atoms, which change them to their reduced forms, NADH and $FADH_2$. Next, NADH and $FADH_2$ donate the electrons to the process of oxidative phosphorylation, which transfers energy to ATP.

TIP

NAD$^+$ and FAD act like electron shuttle buses for the cell. The empty buses, NAD^+ and FAD, drive up to oxidation reactions and collect electron passengers. When the electrons get on the bus, the driver puts up the *H* sign to show that the bus is full. Then the full buses, NADH and $FADH_2$, drive over to reactions that need electrons and let the passengers off. The buses are now empty again, so they drive back to another oxidation reaction to collect new passengers. During cellular respiration, the electron shuttle buses drive a loop between the reactions of glycolysis and the Krebs cycle (where they pick up passengers) to the electron transport chain (where they drop off passengers).

>> **Release carbon dioxide (CO_2):** Cells return CO_2 to the environment as waste, which is great for the autotrophs that require CO_2 to produce the food that heterotrophs eat. (See how it's all connected?)

Different kinds of food molecules enter cellular respiration at different points in the pathway. Cells break down simple sugars, such as glucose, in the first pathway: glycolysis. Cells use the second pathway, the Krebs cycle, for breaking down fatty acids and amino acids.

Following is a summary of how different molecules break down in the first two pathways of cellular respiration:

>> During glycolysis, glucose breaks down into two molecules of pyruvate. The backbone of glucose has six carbon atoms, whereas the backbone of pyruvate has three carbon atoms. During glycolysis, energy transfers result in a net gain of two ATP and two molecules of the reduced form of the coenzyme NADH.

>> Pyruvate is converted to acetyl-coA, which has two carbon atoms in its backbone. One carbon atom from pyruvate is released from the cell as CO_2. For every glucose molecule broken down by glycolysis and the Krebs cycle, six CO_2 molecules leave the cell as waste. (The conversion of pyruvate to acetyl-coA produces two molecules of carbon dioxide, and the Krebs cycle produces four.)

>> During the Krebs cycle, acetyl-coA breaks down into carbon dioxide (CO_2). The conversion of pyruvate to acetyl-coA produces two molecules of NADH. Energy transfers during the Krebs cycle produce an additional six molecules of NADH, two molecules of $FADH_2$, and two molecules of ATP.

Transferring energy to ATP

In the inner membranes of the mitochondria in your cells, hundreds of little cellular machines are busily working to transfer energy from food molecules to ATP. The cellular machines are called *electron transport chains*, and they're made of a team of proteins that sits in the membranes transferring energy and electrons throughout the machines.

REMEMBER

The coenzymes NADH and $FADH_2$ carry energy and electrons from glycolysis and the Krebs cycle to the electron transport chain. The coenzymes transfer the electrons to the proteins of the electron transport chain, which pass the electrons down the chain. Oxygen collects the electrons at the end of the chain. (If you didn't have oxygen around at the end of the chain to collect the electrons, no energy transfer could occur.) When oxygen accepts the electrons, it also picks up protons (H^+) and becomes water (H_2O).

TIP

The proteins of the electron transport chain are like a bucket brigade that works by one person dumping a bucket full of water into the next person's bucket. The buckets are the proteins, or electron carriers, and the water inside the buckets represents the electrons. The electrons get passed from protein to protein until they reach the end of the chain.

While electrons are transferred along the electron transport chain, the proteins use energy to move protons (H^+) across the inner membranes of the mitochondria. They pile the protons up like water behind the "dam" of the inner membranes. These protons then flow back across the mitochondria's membranes through a protein called *ATP synthase* that transforms the kinetic energy from the moving protons into chemical energy in ATP by capturing the energy in chemical bonds as it adds phosphate molecules to ADP.

The entire process of how ATP is made at the electron transport chain is called the *chemiosmotic theory of oxidative phosphorylation* and is illustrated in Figure 4-4.

REMEMBER

At the end of the entire process of cellular respiration, the energy transferred from glucose is stored in 36 to 38 molecules of ATP, which are available to be used for cellular work. (And boy do they get used quickly!)

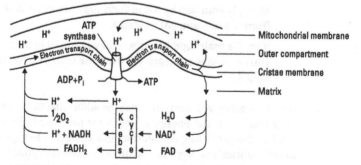

FIGURE 4-4: The events happening inside mitochondria, as described by the chemiosmotic theory.

Energy and Your Body

Your body takes in chemical potential energy when you eat food and then transfers the energy from that food to your cells. As you use the energy to do work, that energy is eventually transformed into heat energy that you transfer to your surroundings.

Energy can be measured in many different ways, but the energy in food is measured in calories. Basically, a *calorie* is a unit of measurement for heat energy. It takes 1 calorie to raise the temperature of 1 gram of water by 1 degree Celsius (*not* Fahrenheit). The calories that you count and see written on food packages are really *kilocalories*. (*Kilo* means "1,000," so a *kilocalorie* is equal to 1,000 calories.) Kilocalories are represented by a capital C, whereas calories are represented by a lowercase c. From here on out, we use the term Calorie (with a capital C) to represent the kilocalories you're familiar with from nutrition facts labels.

You can get an approximate measure of your basic energy needs by performing a simple calculation to determine your *basal metabolic rate* (BMR), the approximate number of Calories you need just to maintain your body's minimum level of activity (breathing, blood pumping, digestion, and so on). Here's how to calculate BMR:

1. **Multiply your weight in pounds by 10.**
2. **Multiply your height in inches by 6.25.**
3. **Add these two numbers together.**
4. **Multiply your age by 5 and then subtract this number from the one you got in Step 3.**
5. **If you're male, add 5 to the total you found in Step 4; if you're female, subtract 161 from the total you found in Step 4.**

If you exercise, you need to consume additional Calories to supply your body with the energy it needs for increased physical activity. Use the preceding calculation and Table 4-1 to figure out how many Calories you need to consume to maintain your lifestyle.

TABLE 4-1 Determining Caloric Need Based on Lifestyle

If You're . . .	Multiply Your BMR by . . .
Fairly sedentary (little or no exercise and desk job)	1.2
Lightly active (light exercise or sports 1 to 3 days per week)	1.375
Moderately active (moderate exercise or sports 3 to 5 days per week)	1.55
Very active (hard exercise or sports 6 to 7 days per week)	1.725
Extremely active (hard daily exercise or sports and physical job)	1.9

In the past, humans had to work hard to find their food and sometimes came up empty-handed. To survive, the human body developed a mechanism for storing energy that can be used during times of low food intake. It packs energy-rich fat onto your hips, thighs, abdomen, and buttocks. So if you take in more Calories in a day than you need, the extra Calories are stored as fat in your adipose tissue. Every 3,500 extra Calories equals 1 pound of fat. And your body doesn't give up extra potential energy easily! If you continue to take in more Calories than you use, you *will* gain weight because it's much easier for your body to create fat than to use it.

how DNA replicates itself

» Discovering how mitosis produces exact copies of cells

» Making egg and sperm cells through meiosis

» Appreciating the power of genetic diversity

Chapter **5**

Reproducing Cells

All living things can reproduce their cells for growth, repair, and reproduction. Asexual reproduction by mitosis creates cells that are genetically identical to the parent cell. Sexual reproduction requires a special type of cell division called meiosis that produces cells containing half the genetic information of the parent cell. Meiosis and sexual reproduction result in greater genetic diversity in offspring and, consequently, in the populations of living things.

In this chapter, we explore the reasons that cells divide and present the steps of each type of cell division. We also introduce you to the ways sexual reproduction adds variety to species of all kinds.

Reproduction: Keep On Keepin' On

Biology is all about life. And, when you think about it, life is really all about continuation — living things keep on keepin' on from one generation to the next, passing on critical genetic information. Certainly this is one of the core differences between organisms and inanimate objects. After all, have you ever seen a chair or table replicate itself? Only living things have the ability to pass on genetic information and replicate themselves.

When cells replicate, they make copies of all their parts, including their DNA, and then divide themselves to make new cells. If a cell makes an exact copy of itself, it's engaging in *asexual reproduction*. Single-celled prokaryotes, such as bacteria, reproduce asexually by binary fission; they're able to divide quickly and reproduce themselves in as little as 10 to 20 minutes. Some single-celled eukaryotes and individual cells within a multicellular eukaryote also reproduce asexually. However, they use a process called mitosis (which we explain in the later section "Mitosis: One for you, and one for you") to produce new generations. If a cell produces a new cell that contains only half of its genetic information, that cell has engaged in *sexual reproduction*. A special type of cell division known as meiosis (which we explain in the later section "Meiosis: It's all about sex, baby") is responsible for all sexual reproduction.

REMEMBER

Cells divide for the following important reasons:

» **To make copies of cells for growth:** You started out as a single cell after mom's egg met dad's sperm, but today you have about 10 trillion cells in your body. All those cells were produced from that first cell and its descendents by mitosis. When you watch plants grow taller or baby animals grow into adults, you're seeing mitosis at work.

» **To make copies of cells for repair:** It's a fact of life that cells wear out and need to be replaced. For instance, you constantly shed skin cells from your surface. If your body couldn't replace these cells, you'd run out of skin. And if an organism gets injured, its body uses mitosis to produce the cells necessary to repair the injury.

» **To carry on the species:** During asexual reproduction, organisms make exact copies of themselves for the purpose of creating offspring. During sexual reproduction, *gametes* (cells, specifically eggs and sperm, containing half the genetic information of their parent cells) get together to make new individuals. When the genetic information of the gametes joins together, the new individual has the correct total amount of DNA.

How DNA Replication Works

If one cell is going to divide to produce two new cells, the first cell must copy all its parts before it can split in half. The cell grows, makes more organelles (see Chapter 3 for the full scoop on organelles), and copies its genetic information (the DNA) so that the new cells each have a copy of everything they need. Cells use a process called *DNA replication* to copy their genetic material. In this process, the original DNA strands serve as the *template* (or guide) for the construction of the new strands. It's particularly important that each new cell receives an accurate copy of the genetic information because this copy, whether it's accurate or faulty, directs the structure and function of the new cells.

The basic steps of DNA replication are as follows:

REMEMBER

1. **The two parental DNA strands separate so that the rungs of the double helix ladder are split apart with one nucleotide on one side and one nucleotide on the other.** (See Chapter 2 for a depiction of a DNA molecule.) The entire DNA strand doesn't unzip all at once, however. Only part of the original DNA strand opens up at one time. The partly open/partly closed area where the replication is actively happening is called the *replication fork* (this is the Y-shaped area in Figure 5-1).

2. **The enzyme DNA polymerase reads the DNA code on the parental strands and builds new partner strands that are complementary to the original strands.** To build complementary strands, DNA polymerase follows the *base-pairing rules* for the DNA nucleotides: A always pairs with T, and C always pairs with G (see Chapter 2 for further details about nucleotides). If, for example, the parental strand has an A at a particular location, DNA polymerase puts a T in the new strand of complementary DNA it's building. When DNA polymerase is done creating complementary pairs, each parental strand has a brand-new partner strand.

DNA polymerase is considered *semiconservative* because each new DNA molecule is half old (the parental strand) and half new (the complementary strand).

REMEMBER

Several enzymes help DNA polymerase with the process of DNA replication (you can see them and DNA polymerase hard at work in Figure 5-1):

>> **Helicase** separates the original parental strands to open the DNA.

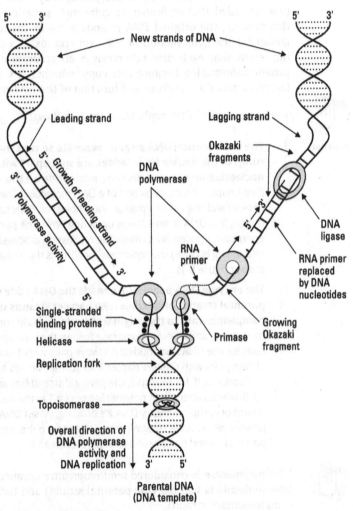

New strands of DNA

Leading strand

Lagging strand

Okazaki
fragments

DNA
polymerase

5′   3′
Growth of leading strand
Polymerase activity
3′
5′

5′   3′

DNA
ligase

RNA
primer

RNA primer
replaced
by DNA
nucleotides

Single-stranded
binding proteins

Growing
Okazaki
fragment

Helicase

Primase

Replication fork

Topoisomerase

Overall direction of
DNA polymerase
activity and
DNA replication   ↓ 3′   5′

Parental DNA
(DNA template)

FIGURE 5-1: DNA replication.

- » **Primase** puts down short pieces of RNA, called *primers,* that are complementary to the parental DNA. DNA polymerase needs these primers in order to get started copying the DNA.

- » **DNA polymerase I** removes the RNA primers and replaces them with DNA, so it's slightly different from the DNA polymerase that makes most of the new DNA. (That enzyme is officially called *DNA polymerase III,* but we refer to it simply as *DNA polymerase.*)

- » **DNA ligase** forms covalent bonds in the backbone of the new DNA molecules to seal up the small breaks created by the starting and stopping of new strands.

The parental strands of the double helix are oriented to each other in opposite polarity: Chemically, the ends of each strand of DNA are different from each other, and the two strands of the double helix are flipped upside down relative to one another. Note in Figure 5-1 the numbers 5' and 3' (read "5 prime" and "3 prime"). These numbers indicate the chemical differences of the two ends. You can see that the 5' end of one strand lines up with the 3' end of the other strand. The two strands of DNA have to be flipped relative to each other in order for the bases that make up the rungs of the ladder to fit together the right way for hydrogen bonds to form between them. Because the two strands have opposite polarity, they're *antiparallel strands.*

REMEMBER

The antiparallel strands of the parent DNA create some problems for DNA polymerase. One quirk of DNA polymerase is that it's a one-way enzyme — it can only make new strands of DNA by lining up the nucleotides a certain way. But DNA polymerase needs to use the parent DNA strands as a pattern, and they're going in opposite directions. As a result, DNA polymerase makes the two new strands of DNA a bit differently from each other, as you can see from the following:

- » **One new strand of DNA, called the *leading strand,* grows in a continuous piece.** Refer to Figure 5-1. See how the new DNA on the left side of the replication fork is growing smoothly? The 3' end of this new strand points toward the replication fork, so after DNA polymerase starts building the new strand, it can just keep going.

>> **One new strand of DNA, called the *lagging strand*, grows in fragments.** Look at Figure 5-1 again. Notice how the right side of the replication fork looks a little messier? That's because the replication process doesn't occur smoothly over here. The 3' end of this new strand points away from the fork. DNA polymerase starts making a piece of this new strand but has to move away from the fork to do so (because it can work only in one direction). DNA polymerase can't go too far from the rest of the enzymes that are working at the fork, however, so it has to keep backing up toward the fork and starting over. As a result, the lagging strand is made in lots of little pieces called *Okazaki fragments.* After DNA polymerase is done making the fragments, the enzyme DNA ligase comes along and forms covalent bonds between all the pieces to make one continuous new strand of complementary DNA.

Cell Division: Out with the Old, In with the New

Cell division is the process by which new cells are formed to replace dead ones, repair damaged tissue, or allow organisms to grow and reproduce. Cells that can divide spend some of their time functioning and some of their time dividing. This alternation between not dividing and dividing is known as the *cell cycle*, and it has specific parts:

>> The nondividing part of the cell cycle is called *interphase*. During interphase, cells are going about their regular business. If the cell is a single-celled organism, it's probably busy finding food and growing. If the cell is part of a multicellular organism, like a human, it's busy doing its job. Maybe it's a skin cell protecting you from bacteria or a fat cell storing energy for later.

>> Cells that receive a signal to divide enter a division process, which is either mitosis or meiosis.

- Cells that reproduce asexually, like a skin cell that needs to replace some of your lost skin, divide by *mitosis,* a process that produces cells that are identical to the parent cell.

- Cells that reproduce sexually enter a special process called *meiosis* that produces special cells called *gametes* (in animals) and *spores* (in plants, fungi, and protists) that have half the genetic information of the parent cell. In you, the only cells that reproduce by meiosis are cells in your gonads. Depending on your gender, your *gonads* are your testes or your ovaries. Cells in testes produce gametes called *sperm,* and cells in ovaries produce gametes called *eggs.*

Mitosis and meiosis have many similarities, but the differences are essential. We cover both processes (as well as the interphase) in the sections that follow, but Table 5-1 can help you sort out the important differences at a glance.

TABLE 5-1   **A Comparison of Mitosis and Meiosis**

Mitosis	Meiosis
One division is all that's necessary to complete the process.	Two separate divisions are necessary to complete the process.
Chromosomes don't get together in pairs.	Homologous chromosomes must be paired up to complete the process, which occurs in prophase I.
Homologous chromosomes don't cross over.	Crossing-over is an important part of meiosis and one that leads to genetic variation.
Sister chromatids separate in anaphase.	Sister chromatids separate only in anaphase II, not anaphase I. (Homologous chromosomes separate in anaphase I.)
Daughter cells have the same number of chromosomes as their parent cells, meaning they're diploid.	Daughter cells have half the number of chromosomes as their parent cells, meaning they're haploid.
Daughter cells have genetic information that's identical to that of their parent cells.	Daughter cells are genetically different from their parent cells.
The function of mitosis is asexual reproduction in some organisms. In many organisms, mitosis functions as a means of growth, replacement of dead cells, and damage repair.	Meiosis creates gametes or spores, the first step in the reproductive process for sexually reproducing organisms, including plants and animals.

Interphase: Getting organized

During interphase, cells engage in the metabolic functions that make them unique. For instance, nerve cells send signals, glandular cells secrete hormones, and muscle cells contract. If cells get a signal to reproduce themselves, they grow, copy all their structures and molecules, and make the structures they need to help cell division proceed in an organized fashion. (*Inter-* means "between," so *interphase* is literally the phase between cell divisions.)

REMEMBER

The nuclear membrane is intact throughout interphase, as you can see in Figure 5-2. The DNA is loosely spread out, and you can't see individual chromosomes. Cells that are going to divide copy their DNA during interphase.

Interphase has three subphases:

>> **G₁ phase:** The G in G₁ stands for *growth*. During this phase, which is typically the longest one of the entire cell cycle, the cell grows and produces cell components. Each chromosome is made up of just a single double-stranded piece of DNA. (*Double-stranded* is just another way of saying that the DNA is a double helix.)

REMEMBER

Some cells actually never leave the G₁ phase. They never divide; instead, they just hang out and do their cellular thing. Human nerve cells are perfect examples of cells that never leave the G₁ phase.

>> **S phase:** The S stands for *synthesis*. This phase is when the cell gets ready to divide and puts the pedal to the metal for DNA replication. Every DNA molecule is copied exactly, forming two *sister chromatids* (a pair of identical DNA molecules) that stay attached to each other in each replicated chromosome. You can see replicated chromosomes in Figure 5-2, in the cell labeled Prophase. Each replicated chromosome looks like an *X*, and each *X* represents two identical sister chromatids held together at a location on the chromosome called the *centromere*.

>> **G₂ phase:** During this phase, the cell is packing its bags and getting ready to hit the road for cell division by making the cytoskeletal proteins it needs to move the chromosomes around. When you look at cells that are dividing, the cytoskeletal proteins look like thin threads, hence their

name — *spindle fibers.* A network of spindle fibers spreads throughout the cell during mitosis to form the mitotic spindle, which is represented by the thin curving lines drawn in the cells in Figure 5-2. The *mitotic spindle* organizes and sorts the chromosomes during mitosis.

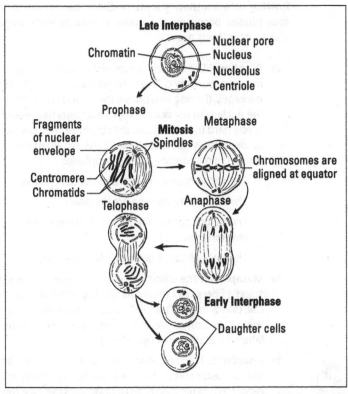

Illustration by Kathryn Born, MA

FIGURE 5-2: Interphase and mitosis.

Mitosis: One for you, and one for you

After interphase is over, cells that are going to divide to create an exact replica of a parent cell enter mitosis, or the M phase of the cell cycle. During mitosis, the cell makes final preparations for its impending split. Processes during mitosis ensure that genetic material is distributed equally so each new daughter cell receives identical information. (Eukaryotic cells are model parents intent on avoiding bickering between their daughter cells.)

The process of mitosis occurs in four phases, with the fourth phase initiating a final process called *cytokinesis*. We outline everything for you in the following sections.

The four phases of mitosis

Although the cell cycle is a continuous process, with one stage flowing into another, scientists divide the events of mitosis into four phases based on the major events in each stage. The four phases of mitosis are

>> **Prophase:** The chromosomes of the cell get ready to be moved around by coiling themselves up into tight little packages. (During interphase, the DNA is spread throughout the nucleus of the cell in long thin strands that would be pretty hard to sort out.) As the chromosomes coil up, or *condense,* they become visible to the eye when viewed through a microscope. During prophase:

- The chromosomes coil up and become visible.
- The nuclear membrane breaks down.
- The mitotic spindle forms and attaches to the chromosomes.
- The nucleoli break down and become invisible.

>> **Metaphase:** The chromosomes are tugged by the mitotic spindle fibers until they're all lined up in the middle of the cell. (*Meta-* means "middle," so it's officially metaphase when the chromosomes are lined up in the middle; see the cell labeled Metaphase in Figure 5-2.)

REMEMBER

>> **Anaphase:** The replicated chromosomes separate so that the two sister chromatids (identical halves) from each replicated chromosome go to opposite sides (see the cell labeled Anaphase in Figure 5-2). This way each new cell has one copy of each DNA molecule from the parent cell when cell division is over.

>> **Telophase:** The cell gets ready to divide into two by forming new nuclear membranes around the separated sets of chromosomes. The two daughter nuclei each have a copy of every chromosome that was in the parent cell, as you can see in Figure 5-2.

TIP

The events of telophase are essentially the reverse of prophase.

- New nuclear membranes form around the two sets of chromosomes.
- The chromosomes uncoil and spread throughout the nucleus.
- The mitotic spindle breaks down.
- The nucleoli reform and become visible again.

Daughter nuclei go their own way: Cytokinesis

REMEMBER

The last order of cell-division business is to give the new daughter nuclei their own cells through a process called *cytokinesis*. (*Cyto-* means "cell," and *kinesis* means "movement," so *cytokinesis* literally means "moving cells.") Cytokinesis occurs differently in animal and plant cells, as you can see from the following list and Figure 5-3:

>> In animal cells, cytokinesis begins with an indentation, called a *cleavage furrow,* in the center of the cell. Cytoskeletal proteins act like a belt around the cell, contracting down and squeezing the cell in two. (Imagine squeezing a ball of dough at the center until it becomes two balls of dough.)

>> In plant cells, a new cell wall forms at the center of the cell. Because a rigid cell wall is involved, the cell can't be squeezed in two. Instead, vesicles deliver wall material to the center of the cell and then fuse together to form the cell plate. The vesicles are basically little bags made of membrane that carry the wall material, so when they fuse together, their membranes form the plasma membranes of the new cells. The wall material gets dumped between the new membranes, forming the plant's cell walls.

After cytokinesis is complete, the new cells move immediately into the G_1 stage of interphase. No one stops to applaud the great accomplishment of successfully completing the mitosis process, which is really too bad because it's the root of renewal and regeneration.

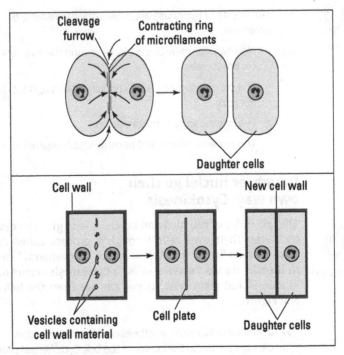

FIGURE 5-3: Cytokinesis.

Meiosis: It's all about sex, baby

Meiosis is unique because the resulting cells have only half their parent cells' *chromosomes,* or singular pieces of DNA. Human body cells have 46 chromosomes, 2 each of 23 different kinds. The 23 pairs of chromosomes can be sorted by their physical similarities and lined up to form a chromosome map called a *karyotype* (see Figure 5-4). The two matched chromosomes in each pair are called *homologous chromosomes.* (*Homo-* means "same," so these are chromosomes that have the same kind of genetic information.) In each pair of your homologous chromosomes, one chromosome came from Mom, and one came from Dad. For every gene that your mom gave you, your dad also gave you a copy, so you have two copies of every gene (with the exception of genes on the X and Y chromosomes if you're male).

REMEMBER

The pairs of homologous chromosomes have the same kind of genetic information. If one of the two has a gene that affects eye color, for example, the other chromosome has the same gene in the same location. The messages in each gene may be slightly different — for example, one gene could have a message for light eyes whereas the other gene could have a message for dark eyes — but both chromosomes have the same type of gene in each location.

1 **2** **3** **4** **5**

6 **7** **8** **9** **10** **11** **12**

13 **14** **15** **16** **17** **18**

19 **20** **21** **22** **X/Y**

Normal Karyotype

FIGURE 5-4: A human karyotype.

Human *gametes* (sperm cells and egg cells) have just 23 chromosomes. Through sexual reproduction (see Figure 5-5), a sperm and an egg join together to create a new individual, returning the chromosome number to 46. If gametes didn't have half the genetic information, then the cell they form together, called a *zygote*, would have twice the normal genetic information for a human. And when gametes are produced, they can't get just any 23 chromosomes — they have to get one of each pair of chromosomes. Otherwise, the zygote would have extras of some chromosomes and be missing others entirely. The resulting person wouldn't have the correct genetic information and probably wouldn't survive.

REMEMBER

Meiosis is the type of cell division that separates chromosomes so gametes receive one of each type of chromosome. In humans, meiosis separates the 23 pairs of chromosomes so that each cell receives just one of each pair. Consequently, gametes have what's known as a *haploid* number of chromosomes, or a single set. When the two gametes unite, they combine their chromosomes to reach the full complement of 46 chromosomes in a normal *diploid* cell (one with a double set of chromosomes, or two of each type).

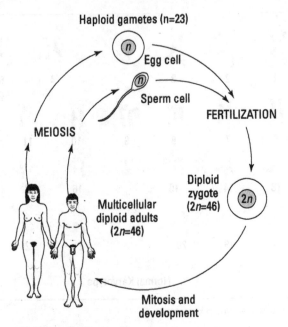

FIGURE 5-5: The human life cycle.

Two cell divisions occur in meiosis, and the two halves of meiosis are called *meiosis I* and *meiosis II*.

>> During meiosis I, homologous chromosomes are paired up and then separated into two daughter cells. Each daughter cell receives one of each chromosome pair, but the chromosomes are still replicated. (Remember that meiosis follows interphase, so DNA replication produced a copy of each chromosome. These two copies, called sister chromatids, are held together, forming replicated chromosomes. You can see that the chromosomes still look like *X*'s after meiosis I in Figure 5-6b.)

>> During meiosis II, the replicated chromosomes send one sister chromatid from each replicated chromosome to new daughter cells. After meiosis II, the four daughter cells each

have one of each chromosome pair, and the chromosomes are no longer replicated. (Notice how the four daughter cells in Figure 5-6b don't have sister chromatids.)

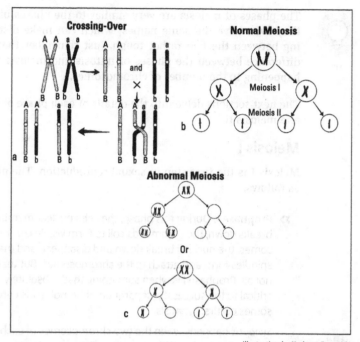

Crossing-Over

Normal Meiosis

and

×

Meiosis I

Meiosis II

b

a

Abnormal Meiosis

Or

c

Illustration by Kathryn Born, MA

FIGURE 5-6: Crossing-over, meiosis, and nondisjunction.

In human males, meiosis takes place after puberty, when the diploid cells in the testes undergo meiosis to become haploid. In females, the process begins a lot earlier — in the fetal stage. While a little girl is dog-paddling around her mother's womb, diploid cells complete the first part of meiosis and then migrate to the ovaries, where they hang out and wait until puberty. With the onset of puberty, the cells take turns entering meiosis II. (Just one per month; no pushing or shoving, please!) Usually one single egg cell is produced per cycle, although exceptions occur, which, if fertilized, lead to fraternal twins, or triplets, or quadruplets . . . you get the idea. The other meiotic cells simply disintegrate.

REMEMBER

When a human sperm cell and a human egg cell — each with 23 chromosomes — unite in the process of fertilization, the diploid condition of the cell is restored. Further divisions by mitosis result in a complete human being.

TIP

The phases of meiosis are very similar to the phases of mitosis; they even have the same names, which can make distinguishing between the two rather tough. Just remember that the key difference between the phases of mitosis and meiosis is what's happening to the number of chromosomes.

The next sections delve into the details of each phase of meiosis I and meiosis II.

Meiosis I

Meiosis I is the first step in sexual reproduction. The phases are as follows:

>> **Prophase I:** During this phase, the cell's nuclear membrane breaks down, the chromatids coil to form visible chromosomes, the nucleoli break down and disappear, and the spindles form and attach to the chromosomes. But that's not all. Prophase I is when something that's absolutely critical to the successful separation of homologous chromosomes occurs: synapsis.

REMEMBER

Synapsis happens when the two chromosomes of each pair find each other and stick together. The synapsis process begins when the homologous chromosomes move and lie next to each other. At this point, the two homologous chromosomes can swap equal amounts of DNA in an event called *crossing-over* (see Figure 5-6a). This swapping of materials results in four completely unique chromatids. This new arrangement of four chromatids is called a *tetrad*.

REMEMBER

Crossing-over between homologous chromosomes during prophase I increases the genetic variability among gametes produced by the same organism. Every time meiosis occurs, crossing-over can happen a little differently, shuffling the genetic deck as gametes are made. This is one of the reasons that siblings can be so different from each other. The process of crossing-over is not always perfect. Crossing-over can sometimes produce duplications or deletions of some genetic material, or crossing-over can take place between two nonhomologous chromosomes. Improper crossing-over can change cell function and sometimes lead to diseases such as cancer.

>> **Metaphase I:** This phase is when the pairs of homologous chromosomes line up in the center of the cell. The difference between metaphase I of meiosis and metaphase of mitosis is that the homologous pairs line up in the former, whereas individual chromosomes line up in the latter.

>> **Anaphase I:** During this phase, the two members of each homologous pair go to opposite sides of the cell, guided by the spindle fibers. Unlike anaphase in mitosis, the sister chromatids remain together and do not separate during anaphase I of meiosis.

>> **Telophase I:** This is when the cell takes a step back (or forward, depending on your perspective) to an interphase-like condition by reversing the events of prophase I. Specifically, the nuclear membrane reforms, the chromosomes uncoil and spread throughout the nucleus, the nucleoli reform, the spindles break down, and the cell splits in two.

Meiosis II

During meiosis II, both daughter cells produced by meiosis I continue their dance of division so that — in most cases — four gametes are the end result. The phases of meiosis II look very similar to the phases of mitosis with one big exception: The cells end up with half the number of chromosomes as the original parent cell.

REMEMBER

Meiosis II separates the sister chromatids of each replicated chromosome and sends them to opposite sides of the cell. Cells going from meiosis I to meiosis II don't go through interphase again (because that would undo all the hard work of meiosis I).

>> **Prophase II:** As in mitosis's prophase and meiosis's prophase I, the nuclear membrane disintegrates, the nucleoli disappear, and the spindles form and attach to the chromosomes.

>> **Metaphase II:** Nothing too exciting here, folks. Just as in any old metaphase, the chromosomes line up at the middle of the cell.

>> **Anaphase II:** The sister chromatids of each replicated chromosome move away from each other to opposite sides of the cell.

>> **Telophase II:** The nuclear membrane and nucleoli reappear, the chromosomes stretch out for the briefest of rests, and the spindles disappear.

After meiosis II, it's time for cytokinesis, which creates four haploid cells (which is impressive considering you had just one diploid cell at the beginning of meiosis).

How Sexual Reproduction Creates Genetic Variation

Sexual reproduction increases genetic variation in offspring, which in turn increases the genetic variability in species. You can see the effects of this genetic variability if you look at the children in a large family and note how each person is unique. Imagine this kind of variability expanded to include all the families you know (not to mention all the families of all the sexually reproducing organisms on Earth), and you begin to get a feel for the dramatic genetic impact of sexual reproduction.

The sections that follow familiarize you with some of the specific causes of genetic variation courtesy of meiosis and sexual reproduction.

Mutations

DNA polymerase occasionally makes uncorrected mistakes when copying a cell's genetic information during DNA replication (which we walk you through earlier in this chapter). These mistakes are called *spontaneous mutations,* and they introduce changes into the genetic code. In addition, exposure of cells to *mutagens* (environmental agents, such as X-rays and certain chemicals, that cause changes in DNA) can increase the number of mutations that occur in cells. When changes occur in a cell that produces gametes, future generations are affected.

Crossing-over

When homologous chromosomes come together during prophase I of meiosis, they exchange bits of DNA with each other. This crossing-over (illustrated in Figure 5-6a) results in new gene combinations and new chances for variety. Crossing-over is one way of explaining how a person can have red hair from his mother's father and a prominent chin from his mother's mother. After crossing-over, these two genes from two different people wound up together on the same chromosome in the person's mother and got handed down together.

Independent assortment

Independent assortment occurs when homologous chromosomes separate during anaphase I of meiosis. When the homologous pairs of chromosomes line up in metaphase I, each pair lines up independently from the other pairs. So, the way the pairs are oriented during meiosis in one cell is different from the way they're oriented in another cell. When the homologous chromosomes separate, many different combinations of homologous chromosomes can travel together toward the same side of the cell. How many different combinations of homologous chromosomes are possible in a human cell undergoing meiosis? Oh, just 2^{23} — that's 8,388,608 to be exact. Now you can begin to see why even large families can have many unique children.

Fertilization

Fertilization presents yet another opportunity for genetic diversity. Imagine millions of genetically different sperm swimming toward an egg. Fertilization is random, so the sperm that wins the race in one fertilization event is going to be different than the sperm that wins the next race. And, of course, each egg is genetically different too. So, fertilization produces random combinations of genetically diverse sperm and eggs, creating virtually unlimited possibilities for variation. That's why every human being who has ever been born — and ever will be born — is genetically unique. Well, almost. Genetically identical twins can develop from the same fertilized egg, but even they can have subtle differences due to development.

Nondisjunction

Nothing's perfect, even in the cellular world, which is why sometimes meiosis doesn't occur quite right. When chromosomes don't separate the way they're supposed to, that's called *nondisjunction*. The point of meiosis is to reduce the number of chromosomes from diploid to haploid, something that normally happens when homologous chromosomes separate from each other during anaphase I. Occasionally, however, a pair of chromosomes finds it just too hard to separate, and both members of the pair end up in the same gamete (see Figure 5-6c).

What happens next isn't pretty. Two of the final four cells resulting from the meiotic process are missing a chromosome as well as the genes that chromosome carries. This condition usually means the cells are doomed to die. Each of the other two cells has an additional chromosome, along with the genetic material it carries. Well, that should be great for these cells, shouldn't it? It should mean they'll have an increased chance for genetic variation, and that's a good thing, right?

Wrong! An extra chromosome is like an extra letter from the IRS. It's not something to hope for. Many times these overendowed cells simply die, and that's the end of the story. But sometimes they survive and go on to become sperm or egg cells. The real tragedy, then, is when an abnormal cell goes on to unite with a normal cell. When that happens, the resulting zygote (and offspring) has three of one kind of chromosome rather than the normal two. The term scientists use for this occurrence is *trisomy*.

Here's the real problem with this scenario: All the cells that develop by mitosis to create the new individual will be trisomic (meaning they'll have that extra chromosome). One possible abnormality occurring from an extra chromosome is *Down syndrome*, a condition that often results in some mental and developmental impairment and premature aging.

Pink and blue chromosomes

Ever wish you could've been born the opposite sex so you wouldn't have to spend so much of your budget on makeup or shave your face every morning? Sorry, but that was never really your decision to make. Like all other genetic characteristics, gender is determined at a chromosomal level.

In many organisms — including humans and fruit flies — the gender of an individual is determined by specific *sex chromosomes*, which scientists refer to as the X and Y chromosomes. The 23 pairs of human chromosomes can be divided into 22 pairs of *autosomes*, chromosomes that aren't involved in the determination of gender, and one pair of sex chromosomes. Men and women have the same types of genes on the 22 autosomes and on the X chromosome. But only guys get a special gene, located on the Y chromosome, that jump-starts the formation of testes in boy fetuses when they're about 6 weeks old. After the testes form, they produce testosterone, and it's usually all boy from then on out. The Y chromosome is smaller than all the other chromosomes, but it packs one powerful little gene!

This Budding's for You: Asexual Reproduction

Asexual reproduction allows organisms to reproduce rapidly and without a partner, which makes asexual organisms essentially just fresher, younger versions of their original selves. Also, asexual organisms don't really die; instead, they just bud off into new versions of themselves and continue on.

REMEMBER

The basic cellular process that makes asexual reproduction possible is *mitosis*, the type of cell division that produces exact copies of parent cells.

Asexual reproduction occurs by several different methods in a variety of animals:

REMEMBER

>> **Budding** happens when a small outgrowth begins on the original organism. That outgrowth gradually becomes larger and eventually separates to create a new individual. Several species of invertebrates, including the hydra, produce offspring by budding.

>> **Fission** occurs when the original organism grows larger and then splits in two. Sea anemones are an example of an invertebrate that reproduces asexually by fission.

>> **Fragmentation** happens when small pieces of the original organism break off and then grow into complete individuals. Starfish are among the animals that use fragmentation to reproduce.

For organisms that are widely separated from others of their kind and for organisms that are doing well in a particular environment, asexual reproduction poses quite the advantage. Yet what makes asexual reproduction an asset for some species — the fact that it doesn't allow for change — also makes it a disadvantage. If a disease strikes or the environment changes and all the organisms are identical, they'll all be affected equally. If the disease can easily kill the organisms, for example, they'll all die. If they were the only organisms of their kind, then an entire species would be wiped out in one fell swoop. Species ultimately have a better chance of surviving changes if their members have some differences from one another.

Chapter **6**

DNA and Proteins: Life Partners

W ithout deoxyribonucleic acid, or DNA for short, your cells — and all the cells of every other living thing on Earth — wouldn't exist. That's because DNA controls the structure and function of organisms, largely because it's essential to the production of the proteins that determine your traits. When changes occur in the DNA of one or more cells, the effects on the organism made up of those cells can be disastrous.

We show you just how important DNA and proteins are to your everyday life in this chapter. Get ready to discover how DNA and RNA (ribonucleic acid) work together to produce proteins, the types of DNA mutations that occur and how they affect you, and more.

Proteins Make Traits Happen, and DNA Makes the Proteins

You probably know that DNA is your genetic blueprint and that it carries the instructions for your traits. But you may not know exactly how your DNA causes you to look and function a certain way. DNA contains the instructions for the construction of the molecules that carry out the functions of your cells. These functional molecules are mostly proteins, and the instructions for creating these proteins are found in your *genes*, sections of DNA that lie along your chromosomes.

TIP

Think of your cells as little factories that have to carry out certain functions. Each function relies upon the actions of the robot workers in the factory. DNA is the instructions for the construction of each type of robot. If the robots are built correctly according to their instructions, they work the way they're supposed to. Based on how the robots work, the factory accomplishes specific tasks. Your cells don't have little robots running around, but they do have worker molecules that carry out the functions of the cell. If one of these molecules isn't working correctly, your cells will work differently than they're supposed to, which could affect your traits.

REMEMBER

One gene equals one blueprint for a functional molecule. Because many of the functional molecules in your cells are proteins, genes often contain the instructions for building the polypeptide chains that make up proteins (for more on protein structure, head to Chapter 2). So, one gene often equals one polypeptide chain.

Sometimes it's hard to imagine how one little protein can affect you in an important way. After all, humans have about 25,000 genes, so you make a lot of different proteins. How can just one faulty protein make a difference? Well, if your skin cells couldn't make the protein collagen, your skin would fall off at the slightest touch. Also, if the cells in your pancreas couldn't make the protein insulin, you'd have diabetes. So you see, all the functions of your body that you probably take for granted — the way it's built, the way it looks, and the way it functions — are controlled by the actions of proteins.

Moving from DNA to RNA to Protein

The instructions in DNA determine the structure and function of all living things, which makes DNA pretty darn important. Every time a cell reproduces, it must make a copy of these instructions for the new cell. When cells need to build a functional molecule (usually a protein), they copy the information in the genes into an RNA molecule instead of using the DNA blueprint directly (see Chapter 3 for more about RNA molecules). Here's an outline of the process:

>> Cells use *transcription* to copy the information in DNA into RNA molecules.

>> The information to build proteins is copied into a special type of RNA called *messenger RNA* (mRNA), which carries the blueprint for the protein from the nucleus to the cytoplasm where it can be used to build the protein.

>> Cells use *translation* to build proteins from the information carried in mRNA molecules.

REMEMBER

The idea that information is stored in DNA, copied into RNA, and then used to build proteins is considered the *central dogma of molecular biology.*

TIP

Transcription and translation are two pretty similar sounding words for two very different processes in cells. One way to remember which process is which is to think about the English meanings of these words. When you transcribe something, you copy it. Transcription in cells takes the information in DNA and uses it to make RNA. DNA and RNA are similar molecules, so it's not like you're really changing anything; you're just copying the information down. When you translate something, on the other hand, you change it from one language to another. Translation in cells takes the information in mRNA and uses it to build a protein, which is a different type of molecule. So, translation changes the language of molecules from RNA to protein.

The sections that follow give you an in-depth look at transcription, RNA processing, and translation.

Rewriting DNA's message: Transcription

DNA molecules are long chains made from four different building blocks called *nucleotides* that biologists represent with the letters A, T, C, and G (flip to Chapter 2 for the scoop on DNA structure).

These chemical units are joined together in different combinations that form the instructions for cells' functional molecules, which are mostly proteins.

When your cells need to build a particular protein, the enzyme RNA polymerase locates the gene for that protein and makes an RNA copy of it. (RNA polymerase is shown in Step 2 of Figure 6-1.) Because RNA and DNA are similar molecules, they can attach to each other just like the two strands of the DNA double helix do. RNA polymerase slides along the gene, matching RNA nucleotides to the DNA nucleotides in the gene.

REMEMBER

The base-pairing rules for matching RNA and DNA nucleotides are almost the same as those for matching DNA with DNA (see Chapter 2). The exception is that RNA contains nucleotides with uracil (U) rather than thymine (T). During transcription, RNA polymerase pairs C with G, G with C, A with T, and U with A. (Figure 6-1 gives you the visual of this. Note that the new RNA strand CAUCCA pairs up with the DNA sequence GTAGGT in the gene.)

It may seem strange that your cells copy the information in your genes into a sort of mirror image made of RNA, but it actually makes a lot of sense. Your genes are extremely important and need to be protected, so they're kept safe in the nucleus at all times. Your cells make copies of any information they need so the original DNA doesn't get damaged.

TIP

You can think of your chromosomes like drawers in a file cabinet. When your cells need information from the files, they open a drawer, take out a file (the gene), and make a copy of the information (the RNA molecule) that can travel out into the world (the cytoplasm). The original blueprint (the DNA) is kept safely locked away in the file cabinet.

Of course, RNA polymerase and DNA aren't the only things involved in transcription. The following sections introduce you to the other players and walk you through the process of transcription step by step.

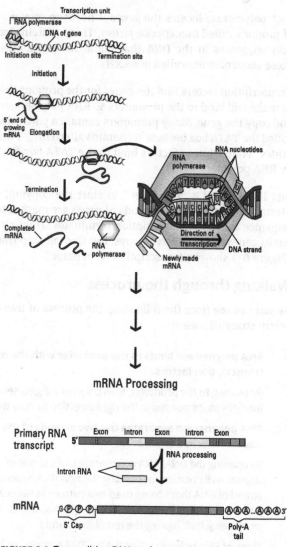

FIGURE 6-1: Transcribing DNA and processing mRNA within the nucleus of an eukaryotic cell.

Finding out what else is involved

RNA polymerase locates the genes it needs to copy with the help of proteins called *transcription factors*. These proteins look for certain sequences in the DNA that mark the beginnings of genes; these sequences are called *promoters*.

Transcription factors find the genes for the proteins the cell needs to make and bind to the promoters so RNA polymerase can attach and copy the gene. Many promoters contain a particular sequence called the TATA box because it contains alternating T and A nucleotides. Transcription factors bind to the TATA box first, followed by RNA polymerase.

Just like we had a "promoter" to start the copying, we have a "terminator" to end it. The ends of genes are marked by a special sequence called the transcription terminator. *Transcription terminators* can work in different ways, but they all stop transcription. (Figure 6-1 shows a transcription terminator.)

Walking through the process

As you can see from the following, the process of transcription is pretty straightforward:

REMEMBER

1. **RNA polymerase binds to the promoter with the help of transcription factors.**

   By binding to the promoter, RNA polymerase gets set up on the DNA so it's pointed in the right direction to copy the gene.

2. **RNA polymerase separates the two strands of the DNA double helix in a small area.**

   By opening the DNA, RNA polymerase can use one of the DNA strands as its pattern for building the new RNA molecule. (The strand of DNA that's being used as a pattern in Figure 6-1 is labeled as the DNA template. The new RNA strand is shown as it's being built against the template strand.)

   Think of RNA polymerase as the "pull" of a zipper. As RNA polymerase slides along the DNA, it opens a new area, and the DNA behind it closes back up.

3. **RNA polymerase uses base-pairing rules to build an RNA strand that's complementary to the DNA in the template strand.**

   Because the base-pairing rules are specific, the new RNA molecule contains a mirror image of the code in the DNA. Remember that in RNA, base T is replaced with U.

4. **RNA polymerase reaches the termination sequence and releases the DNA.**

   Some terminators have a sequence that causes the new RNA to fold up on itself at the end, making a little bump that causes RNA polymerase to get knocked off of the DNA.

REMEMBER

Your cells use transcription to make several types of RNA molecules. Some of these RNA molecules are worker molecules for the cell, others are part of cellular structures, and one type — mRNA — carries the code for proteins to the cytoplasm.

Putting on the finishing touches: RNA processing

After RNA polymerase transcribes one of your genes and produces a molecule of mRNA, the mRNA isn't quite ready to be translated into a protein. In fact, when the mRNA is hot off the presses, it's called a *pre-mRNA* or *primary transcript* because it's not quite finished.

REMEMBER

Before the pre-mRNA can be translated, it has to undergo a few finishing touches via RNA processing (refer to Figure 6-1):

>> **The 5' cap, a protective cap, is added to the beginning of the mRNA.** The *5' cap* tells the cell it should translate this piece of RNA.

>> **The poly-A tail, an extra bit of sequence, is added to the end of the mRNA.** Like its name suggests, the *poly-A tail* is a chain of repeating nucleotides that contain adenine (A). It protects the finished mRNA from being broken down by the cell.

>> **The pre-mRNA is spliced to remove introns (noncoding sequences).** One kind of weird thing about your genes (and ours too) is that the code for proteins is interrupted by sequences called *introns*. Your cells remove the introns before shipping the mRNA out to the cytoplasm.

The sections of the pre-mRNA that wind up getting translated are called *exons*. When your cells cut the introns out of pre-mRNA, the exons all come together to form the blueprint for the protein.

If you get confused about what introns and exons do, just remember that *intron*s *inter*rupt and *exon*s get to *ex*it the nucleus.

TIP

Converting the code to the right language: Translation

After a mature mRNA leaves the nucleus of a cell, it heads for a ribosome in the cell's cytoplasm where the code it contains can be translated to produce a protein (for more on ribosomes, see Chapter 3). As the strand of mRNA slides through the ribosome, the code is read three nucleotides at a time.

A group of three nucleotides in mRNA is called a *codon*. If you take the four kinds of nucleotides in RNA — A, G, C, and U — and make all the possible three-letter combinations you can, you'd come up with 64 possible codons. Each codon specifies one of 20 amino acids in the polypeptide chain of a protein. Some amino acids can be specified by more than one codon.

REMEMBER

To figure out the amino acid a singular codon represents, follow the labels on the edges of the table in Figure 6-2. So to find out what the codon CGU represents:

1. **Look first to the left side of the table and find the row marked by the first letter in the codon.**

   The letter C is the second letter down, so the amino acid represented by the C portion of the codon CGU is found in the second row of the table.

2. **Look to the top of the table and find the column marked by the second letter in the codon.**

   The letter G is the last letter in the row, so the amino acid represented by the G part of the codon CGU is found at the intersection of the second row (which is marked by C) and the last column under the Second Letter heading.

3. **Look to the right side of the table and find the row marked by the third letter in the codon.**

The letter U is listed first, so the amino acid represented by the U portion of the codon CGU is the first one listed in the intersection of the second row and the last column under the Second Letter heading. Put it all together and you find that the amino acid represented by the codon CGU is arginine.

First Letter	Second Letter				Third Letter
↓	U	C	A	G	↓
U	phenylalanine	serine	tyrosine	cysteine	U
	phenylalanine	serine	tyrosine	cysteine	C
	leucine	serine	STOP	STOP	A
	leucine	serine	STOP	tryptophan	G
C	leucine	proline	histidine	arginine	U
	leucine	proline	histidine	arginine	C
	leucine	proline	glutamine	arginine	A
	leucine	proline	glutamine	arginine	G
A	isoleucine	threonine	asparagine	serine	U
	isoleucine	threonine	asparagine	serine	C
	isoleucine	threonine	lysine	arginine	A
	methionine & START	threonine	lysine	arginine	G
G	valine	alanine	aspartate	glycine	U
	valine	alanine	aspartate	glycine	C
	valine	alanine	glutamate	glycine	A
	valine	alanine	glutamate	glycine	G

FIGURE 6-2: The genetic code.

REMEMBER

To translate a molecule of mRNA, begin at the start codon closest to the 5′ cap of the mRNA, divide the message up into codons, and look the codons up in a table of the genetic code that shows the names of the 20 different amino acids found in the proteins of living things. For example, 5′CCGCAUGCGAAAAUGA3′ translates into methionine–arginine–lysine.

The following sections get you acquainted with specialized codons and the anticodons all codons need to pair up with for translation to occur. They also help you understand the overall process of translation.

Making sense of codons and anticodons

The genetic code is amazingly similar for all the organisms on Earth, from you to *E. coli*. To read it, you need to know about the unique features of some of the codons:

>> **The codon AUG is the start codon.** AUG is called the *start codon* because translation begins here. When one of your cells starts translating mRNA into a polypeptide, the first AUG closest to the 5' cap of the mRNA is the first codon to be read. AUG also represents the amino acid methionine, so methionine is the first amino acid added to the polypeptide chain.

>> **The codons UAA, UAG, and UGA are stop codons.** Translation ends when a stop codon is read in the mRNA. A *stop codon* only indicates when translation should end; it doesn't represent an amino acid. When stop codons are read in the mRNA, translation stops without adding any new amino acids to the polypeptide chain.

>> **Some amino acids are represented by more than one codon.** For instance, arginine is represented by more than one codon; CGU, CGC, CGA, and CGG all represent arginine. Because of this situation, biologists say that the genetic code is *redundant* (more than one codon represents some amino acids).

In order for your cells to decode mRNA, they need the help of an important worker: transfer RNA (tRNA). *Transfer RNA* transfer the correct amino acid to the ribosome in order to make the right polypeptide sequence. Like all RNA molecules, tRNA is made of nucleotides that can pair up with other nucleotides according to base-pairing rules.

During translation, tRNA molecules pair up with the codons in mRNA to figure out which amino acid should be added to the chain. Each tRNA has a special group of three nucleotides, called an *anticodon*, that pairs up with the codons in mRNA. Each tRNA also carries a specific amino acid. So, the tRNA that has the right anticodon to pair with a specific codon adds its amino acid to the growing polypeptide chain.

REMEMBER

Because the pairing of anticodon to codon is specific, only one tRNA can pair up with each codon. The specific relationship between tRNA anticodons and mRNA codons ensures that each codon always specifies a particular amino acid.

Breaking down the translation process

Although the process of translation is fairly complicated, it's pretty easy to understand when you break it down into three main steps: the beginning *(initiation)*, the middle *(elongation)*, and the end *(termination)*. Follow along in Figure 6-3 as we present these three steps:

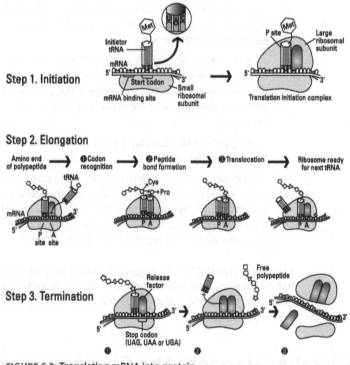

FIGURE 6-3: Translating mRNA into protein.

1. **During initiation, the ribosome and the first tRNA attach to the mRNA (see Step 1 in Figure 6-3).**

   The small subunit of the ribosome binds to the mRNA. Then the first tRNA, which carries the amino acid methionine, attaches to the start codon. The start codon is AUG, so the first tRNA has the anticodon UAC (see Step 1 in Figure 6-3). After the first tRNA is bound to the mRNA, the large subunit of the ribosome attaches to form a complete ribosome.

2. **During elongation, tRNAs enter the ribosome and donate their amino acids to the growing polypeptide chain.**

   Each tRNA enters a pocket in the ribosome called the *A site* (see Step 2 in Figure 6-3). An adjacent pocket, called the *P site,* holds a tRNA with the growing polypeptide chain (see Step 2 in Figure 6-3). When a tRNA is parked in the A site and the P site, the ribosome catalyzes the formation of a *peptide bond* between the growing polypeptide chain and the new amino acid. In Figure 6-3, a bond is forming between the amino acids cysteine (cys) and proline (pro) because they're next to each other in the ribosome.

   After the new amino acid is added to the growing chain, the ribosome slides down the mRNA, moving a new codon into the A site. After a new codon is in the A site, another tRNA can enter the ribosome, and the process of elongation can continue.

3. **During termination, a stop codon in the A site causes translation to end.**

   The ribosome slides down the mRNA until a stop codon enters the A site. When a stop codon is in the A site, an enzyme called a *release factor* enters the ribosome and cuts the polypeptide chain free. Translation stops, and the ribosome and mRNA separate from each other.

Following translation, polypeptide chains may be modified a bit before they fold up and become functional proteins. Often, more than one polypeptide chain combines with another chain to form the complete protein.

Mistakes Happen: The Consequences of Mutation

If a mistake in a strand of DNA goes undetected or unrepaired, the mistake becomes a mutation. A *mutation* is a change from the original DNA strand — in other words, the nucleotides aren't in the order that they should be.

REMEMBER

Mutations in DNA lead to changes in RNA, which can lead to changes in proteins. When proteins change, the functions of cells, and the traits of organisms, can also change.

Mutations usually happen as the DNA is being copied during DNA replication (see Chapter 5 for a description of DNA replication). Two main types of mutations occur:

» **Spontaneous mutations:** These result from uncorrected mistakes by *DNA polymerase,* the enzyme that copies DNA. DNA polymerase is a very accurate enzyme, but it's not perfect. In general, DNA polymerase makes one mistake for every billion base pairs of DNA it copies. One in a billion isn't bad . . . unless you're talking about your DNA. Then, any change can eventually cause problems. Cancer, for example, usually occurs as people age because they've lived long enough to accumulate mutations in certain genes that control cell division.

» **Induced mutations:** These result from environmental agents that increase the error rate of DNA polymerase. Anything that increases the error rate of DNA polymerase is a *mutagen.* The most common mutagens are certain chemicals (such as formaldehyde and compounds in cigarette smoke) and radiation (like ultraviolet light and X-rays).

When mutations occur during DNA replication, some daughter cells formed by mitosis or meiosis inherit the genetic change (we explain how cells divide in Chapter 5). The types of mutations these cells inherit can be divided into three major categories:

» **Base substitutions:** When the wrong nucleotides are paired together in the parent DNA, a *base substitution* occurs. If the parent DNA molecule has a nucleotide containing thymine (T), then DNA polymerase should bring in a nucleotide containing adenine (A) for the new strand of DNA. However, if DNA polymerase makes a mistake and brings in a nucleotide with guanine (G) by mistake, that's a base substitution.

Because just one nucleotide was changed, the mutation is called a *point mutation*. The effect of point mutations ranges from nothing to severe:

- *Silent mutations* have no effect on the protein or organism. Because the genetic code is redundant, changes in DNA may lead to changes in mRNA that don't cause changes in the protein. (See the earlier section "Making sense of codons and anticodons" for more on the redundancy of the genetic code.)

- *Missense mutations* change the amino acids in the protein. Changes in DNA can change the codons in mRNA, leading to the addition of different amino acids into a polypeptide chain. The severity of missense mutations depends on how different the original amino acid is from the new amino acid and where in the protein the change occurs.

- *Nonsense mutations* introduce a stop codon into the mRNA, preventing the protein from being made. If the DNA changes so that a codon in the mRNA becomes a stop codon, then the polypeptide chain gets cut short. Nonsense mutations usually have severe effects and are the cause of many genetic diseases, including certain forms of cystic fibrosis, Duchenne muscular dystrophy, and thalassemia (an inherited form of anemia).

» **Deletions:** When DNA polymerase fails to copy all the DNA in the parent strand, that's a *deletion*. If nucleotides in the parent DNA are read but the complementary bases aren't inserted, the new strand of DNA is missing nucleotides. If one or two nucleotides are deleted, then the codons in the mRNA will be skewed from their proper threes, and the resulting polypeptide chain will be very altered. Mutations that change the way in which the codons are read are called *frameshift mutations,* because they shift the reading frame. Deletions of three nucleotides result in the deletion of one amino acid. Serious diseases such as cystic fibrosis and Duchenne muscular dystrophy result from deletions.

» **Insertions:** When DNA polymerase slips and copies nucleotides in the parent DNA more than once, an *insertion* occurs. Just like deletions, insertions of one or two nucleotides can cause frameshift mutations that greatly alter the polypeptide chain. *Huntington's disease,* an illness that causes the nervous system to degenerate starting when a person is in his 30s or 40s, is caused by insertions of the sequence

CAG into a normal gene up to 100 times. Although the sequence is a multiple of three (so technically not a frame-shift mutation), the abundance of these insertions messes up the reading of the normal genetic code, causing either abnormal protein production or a lack of protein production.

Giving Cells Some Control: Gene Regulation

Even though your DNA is in control of the proteins your body makes, and those proteins are in charge of determining your traits, your cells do have some say in life. Because each one of your cells has a complete set of your chromosomes, your cells are able to practice *gene regulation*, meaning they can choose which genes to use (or not use) and when.

REMEMBER

When a cell uses a gene to make a functional molecule, that gene is *expressed* in the cell. Gene regulation is the process cells use to choose which genes to express at any one time. (Scientists talk about gene regulation as cells turning genes "on" or "off.")

Genes are regulated by the action of proteins that bind to DNA and either help or block RNA polymerase from accessing the genes. In your cells, proteins that help RNA polymerase bind to your genes are called *transcription factors*. They bind to special sequences on the DNA near genes' promoters and make it possible for RNA polymerase to bind to the promoter. Transcription and translation occur, producing the protein in the cell.

Gene regulation allows your cells to do two things: adapt to environmental changes and make it so that each cell type has a distinct role in the body. We fill you in on both in the next sections.

Adapting to environmental changes

The world around you is always changing, which means you need to be able to respond to environmental signals in order to maintain your physiological balance. Gene regulation allows you to do just that. When your cells need to respond to environmental changes, they turn genes on or off to make the proteins needed for the response.

Suppose you're getting too much sunlight. To protect your skin, the cells on the tip of your nose need to darken a bit by making more of the skin pigment melanin. The extra sunlight on your skin triggers certain proteins to bind to the genes needed for melanin production and help RNA polymerase access the genes. RNA polymerase reads the genes, making mRNA that contains the blueprints for the necessary proteins. The mRNA is translated, and the proteins are made. The proteins do their jobs, and the skin on your nose turns a darker color. This example of how your skin gets darker illustrates how cells can access genes when they need them in order to respond to signals from the environment.

Becoming specialists through differentiation

You have more than 200 different types of cells in your body, including skin cells, muscle cells, and kidney cells. Each of these cells does a different job for your body, and like any good craftsman, each of these cell types requires the right tools for its job. To a cell, the right tool for the job is usually a specific protein. For instance, skin cells need lots of the protein keratin, muscle cells needs lots of contractile proteins, and kidney cells need water-transport proteins.

REMEMBER

Cell differentiation is the process that makes cells specialized for certain tasks. Your differentiated cells have all the blueprints for all the tools because they each have a full set of your chromosomes; what makes them different from one another is which blueprint they use.

Cells differentiate from one another due to gene regulation. For example, when a sperm met an egg to form the cell that would become the future you, that first cell had the ability to divide and form all the different cell types your body needed. As that cell and its descendents divided, however, signals caused different groups of cells to change their gene expression. Proteins in these cells bound to the DNA molecules, activating some genes and silencing others. As you grew and developed in your mother's uterus, your cells became more and more distinct from each other. Some cells became part of your nervous tissue, whereas others formed your digestive tract. Each of these changes occurred as cells transcribed and translated the genes for the proteins that they needed to do their particular job.

IN THIS CHAPTER

» Discovering how organisms interact
with each other and their environment

» Analyzing populations and their growth
(or lack thereof)

» Tracing the path of energy and matter
on Earth

Chapter **7**

Ecosystems and Populations

O
ne of the amazing things about this planet is that even
though different parts of the world have different climates,
the organisms living there somehow manage to get what
they need to survive from each other and the world around them.
This chapter explores Earth's various ecosystems and details how
the interactions among organisms work to keep life on Earth in
balance. It also covers how scientists study groups of organisms to
stay on top of how their populations are growing (or declining).

Ecosystems Bring It All Together

Life thrives in every environment on Earth, and each of those
environments is its own *ecosystem*, a group of living and nonliving
things that interact with each other in a particular environment.
An ecosystem is essentially a little machine made up of living and
nonliving parts. The living parts, called *biotic factors*, are all the
organisms that live in the area. The nonliving parts, called *abiotic
factors*, are the nonliving things in the area (think air, sunlight,
water, and soil).

Ecosystems exist in the world's oceans, rivers, forests, and they even exist in your backyard and local park. They can be as huge as the Amazon rain forest or as small as a rotting log. The catch is that the larger an ecosystem is, the greater the number of smaller ecosystems existing within it. For example, the ecosystem of the Amazon rain forest also consists of the soil ecosystem and the cloud forest ecosystem (found at the tops of the trees).

REMEMBER

A particular branch of science called *ecology* is devoted to the study of ecosystems, specifically how organisms interact with each other and their environment. Scientists who work in this branch are called *ecologists*, and they look at the interactions between living things and their environment on many different scales, from large to small.

The sections that follow explain how ecologists classify Earth's various ecosystems and how they describe the interactions among the planet's many species. Before you check them out, take a look at Figure 7-1 to get an idea of how living things are organized.

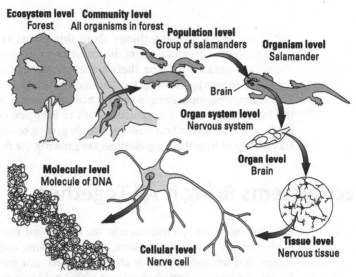

FIGURE 7-1: The organization of living things.

Biomes: Communities of life

All the living things together in an ecosystem form a *community*. For example, a forest community may contain trees, shrubs, wildflowers, squirrels, birds, bats, insects, mushrooms, bacteria, and much more. The different types of communities found on Earth are called *biomes*. Six major types of biomes exist:

>> **Freshwater biomes** include ponds, rivers, streams, lakes, and wetlands. Only about 3 percent of the Earth's surface is made up of freshwater, but freshwater biomes are home to many different species, including plants, algae, fish, and insects.

>> **Marine biomes** contain saltwater and include the oceans, coral reefs, and estuaries. They cover 75 percent of the Earth's surface and are very important to the planet's oxygen and food supply — more than half the photosynthesis that occurs on Earth occurs in the ocean.

>> **Desert biomes** receive minimal amounts of rainfall and cover approximately 20 percent of the Earth's surface. Plants and animals that live in deserts have special adaptations, such as the ability to store water or only grow during the rainy season, to help them survive in the low-water environment.

>> **Forest biomes** contain many trees or other woody vegetation; cover about 30 percent of the Earth's surface; and are home to many different plants and animals, including trees, skunks, squirrels, wolves, bears, birds, and wildcats.

>> **Grassland biomes** are dominated by grasses, but they're also home to many other species, such as birds, zebras, giraffes, lions, buffaloes, termites, and hyenas. Grasslands cover about 30 percent of the Earth's surface and are typically flat, have few trees, and possess rich soil.

>> **Tundra biomes** are very cold and have very little liquid water. Most of the water above and below ground is nearly always frozen, creating a condition called *permafrost*. Tundras cover about 15 percent of the planet's surface and are found at the poles of the Earth as well as at high elevations.

Interactions among species

Not all the organisms in a given community are the same. In fact, they're often members of different species (meaning they can't sexually reproduce together). Yet these organisms must interact with each other as they go about their daily business of finding what they need to survive. The term *ecological niche* is used to describe how the species in a given community interact with each other, their specific environment, and the resources in the environment.

Ecologists use a few specific terms to describe the types of interactions among different species:

REMEMBER

>> **Mutualism:** Both organisms benefit in a mutualistic relationship. Case in point: You give the bacteria in your small intestines a nice place to live complete with lots of food, and they make vitamins for you.

>> **Competition:** Both organisms suffer in a competitive relationship. If a resource such as food, space, or water is in limited supply, species struggle with each other to obtain enough to survive. Just think of a vegetable garden that's overrun with weeds.

>> **Predation and parasitism:** One organism benefits at the expense of the other in predatory and parasitic relationships. When a lion eats a gazelle, the benefits are purely the lion's.

Studying Populations

Each group of organisms of the same species living in the same area forms a *population*. For instance, the forests in the Pacific Northwest consist of Douglas fir trees and Western red cedars. Because Doug firs and Western red cedars are two different kinds of trees, ecologists consider two groups of these trees in the same forest to be two different populations.

REMEMBER

Population ecology is the branch of ecology that studies the structures of populations and how they change.

The following sections introduce you to some of the basic concepts of population ecology. They also help you understand the ways in which populations grow and change, as well as how scientists measure and study their growth.

The basics of population ecology

Like all ecologists, population ecologists are interested in the interactions of organisms with each other and with their environment. The unique thing about population ecologists, though, is that they study these relationships by examining the properties of populations rather than individuals.

The next few sections walk you through some of the basic properties of populations and show you why they're important.

Population density

REMEMBER

One way of looking at the structure of a population is in terms of its *population density* (how many organisms occupy a specific area).

Say you want to get an idea of how the human population is distributed in the state of New York. About 19.5 million people live in the 47,214 square miles that make up the state. If you divide the number of people by the area, you get a population density of about 413 people per square mile. However, the human population of New York isn't evenly distributed.

The New York City metropolitan area has 8,214,426 people living in just 303 square miles, creating a population density of 27,110 people per square mile. These numbers show that the human population in New York is heavily concentrated in New York City and much less concentrated in other areas of the state.

Dispersion

REMEMBER

Population ecologists use the term *dispersion* to describe the distribution of a population throughout a certain area. Populations disperse in three main ways:

>> **Clumped dispersion:** In this type of dispersion, most organisms are clustered together with few organisms in between. Examples include people in New York City, bees in a hive, and ants in a hill.

>> **Uniform dispersion:** Uniformly dispersed organisms are spread evenly throughout an area. Grapevines in a vineyard and rows of corn plants in a field are examples of uniform dispersion.

>> **Random dispersion:** In this type of dispersion, one place in the area is as good as any other for finding the organism. (*Note:* Random dispersion is rare in nature but may result when seeds or larvae are scattered by wind or water.) Examples of random dispersion include barnacles scattered on the surfaces of rocks and plants with wind-blown seeds settling down on bare ground.

Population dynamics

REMEMBER

Population dynamics are changes in population density over time or in a particular area. Population ecologists typically use age-structure diagrams to study these changes and note trends.

Age–structure diagrams, sometimes called *population pyramids*, show the numbers of people in each age group in a population at a particular time. The shape of an age–structure diagram can tell you how fast a population is growing.

>> **A pyramid-shaped age-structure diagram indicates the population is growing rapidly.** Take a look at Figure 7-2a. In Mexico, more people are below reproductive age than above reproductive age, giving the age-structure diagram a wide base and a narrow top. The newest generations are larger than the generations before them, so the population size is increasing.

>> **An evenly shaped age-structure diagram indicates the population is relatively stable.** According to Figure 7-2b, the number of people above and below reproductive age in Iceland is about equal, with a decrease in the population as the older group ages.

>> **An age-structure diagram that has a smaller base than middle portion indicates the population is decreasing in size.** When you refer to Figure 7-2c, you notice that more people are above reproductive age in Japan than below it.

Survivorship

Scientists interested in *demography* — the study of birth, death, and movement rates that cause change in populations — noticed that different types of organisms have distinct patterns in how long offspring survive after birth. The scientists followed groups of organisms that were all born at the same time and looked at their *survivorship*, which is the number of organisms in the group that are still alive at different times after birth.

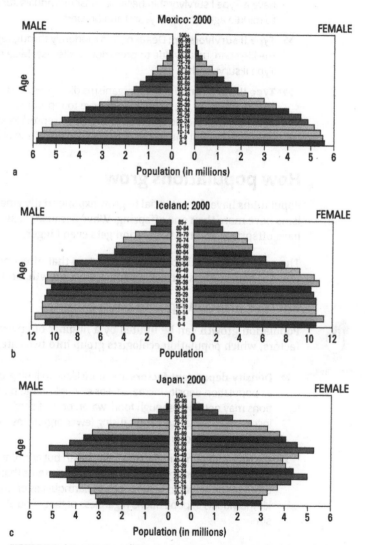

FIGURE 7-2: Age-structure diagrams break down age groups in populations.

Three types of survivorship exist:

>> **Type I survivorship:** Most offspring survive, and organisms live out most of their life span, dying in old age. Humans have a Type I survivorship because most individuals survive to middle age (about 40 years) and beyond.

>> **Type II survivorship:** Death occurs randomly throughout the life span, usually due to predation or disease. Mice have Type II survivorship.

>> **Type III survivorship:** Most organisms die young, and few members of the population survive to reproductive age. However, individuals that do survive to reproductive age often live out the rest of their life span and die in old age.

How populations grow

Populations have the potential to grow exponentially when organisms have more than one offspring. Why? Because those offspring have offspring, and the population gets even bigger.

The next sections fill you in on the factors that affect population growth, how scientists track a population's growth, and more.

Factors affecting population growth

REMEMBER

Population growth can be limited by a number of environmental factors, which population ecologists group into two categories:

>> **Density-dependent factors** are more likely to limit growth as population density increases. For example, large populations may not have enough food, water, or nest sites, increasing competition and causing fewer organisms to survive and reproduce.

>> **Density-independent factors** limit growth but aren't affected by population density. Changes in weather patterns that cause droughts or natural disasters such as earthquakes or floods kill individuals in populations regardless of that population's size.

Some populations can remain very steady in the face of these factors, whereas others fluctuate quite a bit.

>> **Populations that depend on limited resources fluctuate more than populations that have ample resources.** If a population depends heavily on one type of food, for example, and that food becomes unavailable, the death rate will increase rapidly.

>> **Populations with low reproductive rates are more stable than populations with high reproductive rates.** Organisms with high reproductive rates may have sudden booms in population as conditions change. Organisms with low reproductive rates don't experience these booms.

>> **Populations may rise and fall because of interactions between predators and prey.** When prey is abundant, predator populations grow until the increased numbers of predators eat up most of the prey. Then predator numbers fall, allowing prey to recover.

Reaching carrying capacity

REMEMBER

When a population hits *carrying capacity*, it has reached the maximum amount of organisms of a single population that can survive in one *habitat* (the scientific name for a home).

As populations approach the carrying capacity of a particular environment, density-dependent factors have a greater effect, and population growth slows dramatically. If carrying capacity is exceeded even temporarily, the habitat may be damaged, further reducing the amount of resources available and leading to increased deaths.

Taking a closer look at the human population

There's no doubt about it: Humans are the dominant population on Earth, and our numbers keep on rising. It's important to have an understanding of how our population is growing because of the impact humans have on the planet and all the other species on it.

Up until about a thousand years ago, human population growth was very stable. Food wasn't as readily available as it is now. Nor were there antibiotics to fend off invading bacteria, vaccines to fight against deadly diseases, and sewage treatment plants to ensure that water was safe to drink. People didn't shower or wash their hands as often, so they spread diseases more easily. All these factors, and more, increased the death rate and decreased the birth rate of the human population.

Yet in the last 100 to 200 years, the food supply has increased and hygiene and medicines have reduced deaths due to common illnesses and diseases. So not only are more people born, but more of these people are surviving well past middle age. As you can see in Figure 7-3, the human population has grown exponentially in relatively recent history.

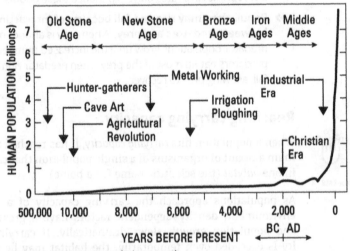

FIGURE 7-3: Human population growth.

At current growth rates, the human population is projected to reach 8 to 12 billion by the end of the 21st century.

What's scary is that scientists question whether the Earth can even support that many humans. The exact carrying capacity of the Earth for humans isn't known because, unlike other species, humans can use technology to increase the Earth's carrying capacity for the species. Currently, scientists estimate that humans are using about 19 percent of the Earth's *primary productivity*, which is the ability of living things like plants to make food. Humans also use about half of the world's freshwater. If humans continue to use more and more of the Earth's resources, the increased competition will drive many other species to extinction. (This pressure on other species from human impacts is already being seen, endangering species such as gorillas, cheetahs, lions, tigers, sharks, and killer whales.)

Moving Energy and Matter around within Ecosystems

Organisms interact with their environment and with other organisms to acquire energy and matter for growth. The interactions among organisms influence behavior and help the organisms establish complex relationships.

REMEMBER

One of the most fundamental ways that organisms interact with each other is eating each other. In fact, all the various organisms in an ecosystem can be divided into four categories called *trophic levels* based on how they get their food:

>> **Producers** make their own food. Plants, algae, and green bacteria are all producers that use energy from the sun to combine carbon dioxide and water and form carbohydrates via photosynthesis.

>> **Primary consumers** eat producers. Because producers are mainly plants, primary consumers are also called *herbivores* (plant-eating animals).

>> **Secondary consumers** eat primary consumers. Because primary consumers are animals, secondary consumers are also called *carnivores* (meat-eating animals).

>> **Tertiary consumers** eat secondary consumers, so they're also considered carnivores.

Organisms in the different trophic levels are linked together in a *food chain*, a sequence of organisms in a community in which each organism feeds on the one below it in the chain.

REMEMBER

Interactions in ecosystems go way beyond a simple food chain because

>> **Some organisms eat at more than one trophic level.** You, for example, may eat a slice of pizza with pepperoni. The grain that made the crust came from a plant, and the pepperoni came from an animal.

>> **Some organisms eat more than one type of food.** Organisms such as humans that eat both plants and animals are called *omnivores*.

>> **Some organisms get their food by breaking down dead things.** *Decomposers*, like bacteria and fungi, release enzymes onto dead organisms, breaking them down into smaller components for absorption.

Organisms that eat more than one type of food belong to more than one food chain. When all the food chains from an ecosystem are put together, they form an interconnected *food web*.

Going with the (energy) flow

The energy living things need to grow flows from one organism to another through food. Sounds simple, we know, but that energy is governed by a few key principles, perhaps most important of which is that an organism never gets to use the full amount of energy it receives from the thing it's "eating."

Energy principles

Some really important energy principles form the foundation of organism interactions in ecosystems:

>> **Energy can't be created or destroyed.** This statement represents a fundamental law of the universe called the *First Law of Thermodynamics*. The consequence of this law is that every living thing has to get its energy from somewhere. No living thing can make the energy it needs all by itself.

>> **When energy is moved from one place to another, it's transferred.** When a primary consumer eats a producer, the energy that was stored in the body of the producer is transferred to the primary consumer.

>> **When energy is changed from one form to another, it's transformed.** When plants do photosynthesis, they absorb light energy from the sun and convert it into the chemical energy stored in carbohydrates. So, during photosynthesis, light energy is transformed into chemical energy.

>> **When energy is transferred in living systems, some of the energy is transformed into heat energy.** After energy is transformed to heat, it's no longer useful as a source of energy to living things. In fact, only about 10 percent of the energy available at one trophic level is usable to the next trophic level.

TIP

Never use the words *lost, disappear, destroyed,* or *created* when you're talking about energy. Use the words *transfer* and *transformed* instead, and you'll avoid a great deal of confusion.

The energy pyramid

Scientists use an *energy pyramid* (also called a *trophic pyramid;* see Figure 7-4) to illustrate the flow of energy from one trophic level to the next. Energy pyramids show the amount of energy at each trophic level in proportion to the next trophic level — what ecologists refer to as *ecological efficiency.* To estimate ecological efficiency, ecologists use what's called the *10-percent rule,* which says that only about 10 percent of the energy available to one trophic level gets transferred to the next trophic level.

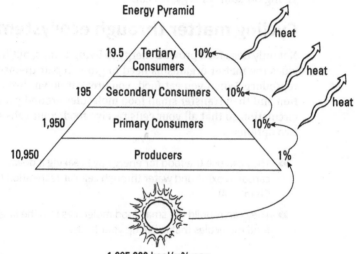

FIGURE 7-4: The energy pyramid.

Following along in Figure 7-4, you see that energy travels to Earth from the sun. About 1 percent of the energy available to producers is captured and stored in food. Producers grow, transferring much of their stored energy into ATP for cellular work and the molecules that make up their bodies. As producers transfer energy for growth, some energy is also transformed into heat that's transferred to the environment.

About 10 percent of the energy that was originally stored in producers is transferred to primary consumers when they eat the producers. Just like producers, the primary consumers grow, transferring energy from food into ATP for cellular work and into the molecules that make up their bodies. As primary consumers transfer energy for growth, some energy is also transformed into heat that's transferred to the environment. This process repeats itself when secondary consumers eat primary consumers and when tertiary consumers eat secondary consumers.

But the energy pyramid doesn't end there. As organisms die, some of their remains become part of the environment. Decomposers and detritivores use this dead matter as their source of food, transferring energy from food into ATP and molecules and giving off some energy as heat.

Cycling matter through ecosystems

Not only does food provide energy to living things, but it also provides the matter organisms need to grow, repair themselves, and reproduce. When you eat food, you break it down through digestion and then transfer small food molecules around your body via circulation so that all your cells receive food. Your cells then have two options:

>> They can use the food for energy by breaking it down into carbon dioxide and water through cellular respiration (see Chapter 4).

>> They can rebuild the small food molecules into the larger food molecules that make up your body.

You don't use food molecules directly to build your cells; you break them down first and use the pieces to build what you need. In other words, your cells are made of human molecules that are rebuilt from the parts of molecules taken from the plants and animals you've eaten.

TIP

Think about what all goes into a slice of pepperoni pizza. The crust came from the grains of plants, and the pepperoni (for the sake of argument) came from a pig. Plants make their own food from carbon dioxide and water and then use that food to build their bodies, which means the plant that went into your pizza crust got the parts it needed to build its body from carbon dioxide in the air and water in the soil. Pigs get their molecules by eating whatever food the farmer gives them, which is likely some type of plant. After you eat a slice of pepperoni pizza, you can trace some of the atoms that make up your body back to carbon dioxide from the air, water from the soil, and plants that were fed to pigs.

REMEMBER

All the carbon, hydrogen, oxygen, nitrogen, and other elements that make up the molecules of living things have been recycled over and over throughout time. Consequently, ecologists say that matter cycles through ecosystems.

Scientists track the recycling of atoms through cycles called *biogeochemical cycles*. Four biogeochemical cycles that are particularly important to living things are the hydrologic cycle, the carbon cycle, the phosphorous cycle, and the nitrogen cycle.

The hydrologic cycle

The *hydrologic cycle* (also known as the *water cycle*) refers to plants obtaining water by absorbing it from the soil and animals obtaining water by drinking it or eating plants or other animals that are made mostly of water. Water returns to the environment when plants transpire and animals perspire. Water evaporates into the air and is carried around the Earth by wind. As moist air rises and cools, water condenses again and returns to the Earth's surface as precipitation (rain, snow, sleet, and hail). Water moves over the Earth's surface in bodies of water such as lakes, rivers, oceans, and even glaciers; it also moves through the groundwater below the soil.

The carbon cycle

The carbon cycle (depicted in Figure 7-5) may be the most important biogeochemical cycle to living things because the proteins, carbohydrates, and fats that make up their bodies all have a carbon backbone. In the carbon cycle, plants take in carbon dioxide from the atmosphere, using it to build carbohydrates via photosynthesis. Animals consume plants or other animals, incorporating the carbon that was in their food molecules into the molecules that make up their own bodies. Decomposers break down dead material, incorporating the carbon from the dead matter into their bodies.

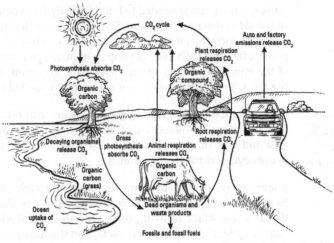

Illustration by Kathryn Born, MA

FIGURE 7-5: The carbon cycle.

All these living things — producers, consumers, and decomposers — also use food molecules as a source of energy, breaking the food molecules back down into carbon dioxide and water in the process of cellular respiration (see Chapter 4).

The phosphorus cycle

Phosphorous is an important component of the molecules that make up living things. It's found in adenosine triphosphate (ATP), the energy-storing molecule produced by every living thing, as well as the backbones of DNA and RNA molecules. The phosphorous cycle involves plants obtaining phosphorous when they absorb inorganic phosphate and water from the soil and animals obtaining phosphorous when they eat plants or other animals. Phosphorus is excreted through the waste products created by animals, and it's released by decomposers back into the soil as they break down dead materials. When phosphorus gets returned to the soil, it's either absorbed again by plants or it becomes part of the sediment layers that eventually form rocks. As rocks erode by the action of water, phosphorus is returned to water and soil.

The nitrogen cycle

Not only is nitrogen part of the amino acids that make up proteins, but it's also found in DNA and RNA. Nitrogen also exists in several inorganic forms in the environment, such as nitrogen gas (in the atmosphere) and ammonia or nitrates (in the soil).

Because nitrogen exists in so many forms, the nitrogen cycle (shown in Figure 7-6) is pretty complex.

> **» Nitrogen fixation occurs when atmospheric nitrogen is changed into a form that's usable by living things.**
> Nitrogen gas in the atmosphere can't be incorporated into the molecules of living things, so all the organisms on Earth depend upon the activity of bacteria that live in the soil.

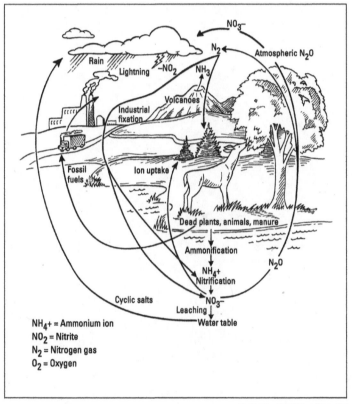

Illustration by Kathryn Born, MA

FIGURE 7-6: The nitrogen cycle.

>> **Ammonification releases ammonia into the soil.** As decomposers break down the proteins in dead things, they release some of it into the soil as ammonia (NH_3).

>> **Nitrification converts ammonia to nitrite and nitrate.** Certain bacteria get their energy by converting ammonia (NH_3) into nitrite (NO_2^-) or nitrite into nitrate (NO_3^-).

>> **Denitrification converts nitrate to nitrite and nitrogen gas.** Some bacteria in the soil use nitrate (NO_3^-) rather than oxygen for cellular respiration. When these bacteria use nitrate, they convert it into nitrite that's released into the soil or nitrogen gas that's released into the atmosphere.

Chapter **8**

Understanding Genetics

G enetics is the branch of biology that looks at how parents pass traits on to their offspring. It all started more than 150 years ago when a monk named Gregor Mendel conducted breeding experiments with pea plants that led him to discover the fundamental rules of inheritance. Although Mendel worked with peas, his ideas explain a lot about why you look and function the way you do.

In this chapter, we show you one of Mendel's experiments and present some of the most important rules of inheritance.

Heritable Traits and the Factors Affecting Them

Dogs have puppies, hens have chicks, and your parents had you. What do all three sets of parents have in common? They all passed their traits on to their offspring. Traits that are inherited from one generation to the next are called *heritable traits.*

When living things reproduce, they make copies of their DNA and pass some of that DNA on to the next generation. The DNA is the genetic code for the traits of the organism. Of course, in sexually reproducing species, offspring aren't exactly like their parents for several reasons:

>> **Offspring receive half their genetic information from their father and half from their mother.** Parents divide their genetic information in half through the process of meiosis (described in Chapter 5).

>> **Even heritable characteristics can change slightly.** DNA changes slightly every time it's copied due to mutation (see Chapter 5). If a mutation is passed from parent to offspring, the offspring may have a new trait.

>> **Some traits are acquired rather than inherited.** Riding a bicycle, speaking French, and swimming are all *acquired traits,* abilities you aren't born with but that you gain during your lifetime.

>> **Some inherited traits are affected by the environment.** The basic color of, say, your skin or hair is written in the code of your DNA, but if you spend lots of time basking in sunlight, your skin will grow darker, and your hair will become lighter.

Mendel's Laws of Inheritance

People have probably always realized that parents pass traits on to their children. After all, as soon as a new baby is born, people start trying to decide who the baby looks like. Yet the first person who really figured out the fundamentals of how traits are passed down was an Austrian monk named Gregor Mendel.

Mendel lived in the mid-19th century. During his time, people believed in *blending inheritance,* meaning they thought that the traits of a father blended with the traits of a mother to produce children whose traits were supposed to be the averages of the parents' traits. So a tall father and a short mother were expected to have average-size kids. Mendel, who was very interested in science and math, tested these ideas about inheritance by breeding pea plants in the abbey garden. He studied many of the heritable traits of peas, including flower color, pea color, plant height, and pea shape. Although other people had bred plants and animals for desirable characteristics before, Mendel was extremely careful in his experiments and used math to look at inheritance in a new way, revealing patterns that no one else had noticed.

Pure breeding the parentals

Mendel used *pure-breeding* plants (plants that always reproduce the same characteristics in their offspring) to ensure that he knew exactly what genetic message he was starting with when he chose a particular plant for his experiments.

To make pure-breeding plants for his experiments, Mendel self-pollinated plants that had the characteristic he wanted to study, weeding out any offspring that were different until all the offspring always had his chosen characteristic. For example, Mendel self-pollinated tall pea plants, pulling out any short offspring until he had plants that would produce only tall offspring. He did the same thing for short pea plants, self-pollinating them until they bred purely short offspring.

REMEMBER

Pure-breeding organisms that are used as the parents in a genetic cross are called *parentals*, or the *P1 generation*.

Analyzing the F1 and F2 generations

In one of his experiments, Mendel bred tall pea plants with short pea plants. According to the idea of blending inheritance, all the offspring should have been average in height. However, when Mendel mated his parentals (called the *P1 generation*) and grew the offspring (called the *F1 generation*), all the offspring were tall. It almost seemed like the short characteristic had disappeared, but when Mendel mated tall pea plants from this new F1 generation and grew their offspring, he saw both tall and short pea plants, indicating that the short characteristic had merely been hidden. The second generation (the *F2 generation*) had about three times as many tall pea plants as short pea plants.

Reviewing Mendel's results

The results of Mendel's pea plant experiments were very exciting because they didn't follow what was expected. In other words, they revealed something new about inheritance.

From his results, Mendel proposed several ideas that laid the foundation for the science of genetics:

>> Traits are determined by factors that are passed from parents to offspring. Today, people call these factors *genes*.

REMEMBER

>> Each organism has two copies of the genes that control every trait. The offspring winds up with two copies of each gene because it gets one copy from Mom and one copy from Dad.

>> Some variations of genes can hide the effects of other variations. Variations that are hidden are *recessive*, whereas variations that hide or mask other variations are *dominant* — this is known as *Mendel's Law of Dominance*. In Mendel's cross between tall and short pea plants, the tall characteristic hid the short characteristic; therefore, the tall gene was the dominant one.

>> The genes that control traits don't blend with each other, nor do they change from one generation to the next. Mendel knew this because the short characteristic, which had disappeared in the F1 generation, reappeared in the F2 generation.

REMEMBER

Sexually reproducing organisms have two copies of every gene, but they give only one copy of each gene to their offspring. Mendel said that this is because the two copies of genes *segregate* (separate from each other) when the organisms reproduce. Scientists now call this idea *Mendel's Law of Segregation*.

So far in this chapter, we have focused on only one gene that controls one trait (plant height), but of course many genes are required to grow a pea plant, and each gene has at least two *alleles* (different forms of the gene, which we discuss in the next section). What Mendel observed was that the alleles for different genes or traits separate out independently of one another — this is referred to as *Mendel's Law of Independent Assortment*. For example, pea shape and height of the pea plant are not dependent on one another.

Defining Key Genetics Terms

Mendel's fundamental ideas have stood the test of time, but geneticists have discovered a great deal more about inheritance since Mendel's day. As the science of genetics grew and developed, so did the language used by geneticists.

REMEMBER

A few key genetics terms are particularly useful when talking about inheritance:

>> **Genes:** Defined as factors that control traits, each *gene* is a section of nucleotides along the chain of DNA in a chromosome. Some genes are thousands of nucleotides long; others are less than a hundred. Your cells have about 25,000 different genes scattered among your 46 chromosomes. Each gene is the blueprint for a worker molecule in your cells, usually a protein. Genes determine protein shape and function, and the actions of proteins control your traits.

>> **Alleles:** Different forms of a gene are called *alleles;* they explain why Mendel saw both tall and short pea plants. Logically, the gene that controls pea plant height has two variations, or alleles — one for tallness and one for shortness. Plants with two alleles for tallness are tall; plants with two alleles for shortness are short. As Mendel saw, plants that have one of each kind of allele are also tall, indicating that the tall allele can hide the effects of the short allele. In other words, the allele for tall is dominant to the allele for short in pea plants; the short allele is this case is called the *recessive* allele.

>> **Loci:** These are the locations on a chromosome where genes are found. Each gene is located at a specific place, or *locus,* on its chromosome.

Many human traits aren't controlled by just one gene. Traits such as your height, weight, and the color of your skin, hair, and eyes result from the interaction of several genes. These traits are called *polygenic traits* (*poly* means "many," and *genic* means "genes"). Polygenic traits usually show a wide range of variety in the population. Pea plants, for example, are either short or tall with nothing in between, whereas adult humans range over a wide variety of heights. This difference is because human height is polygenic and pea height is controlled by just one gene.

Bearing Genetic Crosses

Geneticists use their own unique shorthand when analyzing the results from a *genetic cross* (a mating between two organisms with characteristics that scientists want to study). They use a letter to stand for each gene, capitalizing the letter for dominant alleles. The same letter of the alphabet is used for each allele to show that they're variations of the same gene.

For the cross Mendel did between tall and short pea plants, the letter *T* can be used to represent the gene for plant height. In Figure 8-1, the dominant allele for tallness is shown as *T*, whereas the recessive allele for shortness is shown as *t*.

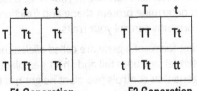

F1 Generation   **F2 Generation**

FIGURE 8-1: Punnett squares showing Mendel's cross between tall and short pea plants.

Geneticists also have special terms for describing the organisms involved in a genetic cross. Here they are:

>> **Genotype:** The combination of alleles that an organism has is its *genotype*. The genotypes of the two parental plants shown in Figure 8-1 are *TT* and *tt*.

>> **Phenotype:** The appearance of an organism's traits is its *phenotype*. The phenotypes of the two parental plants shown in Figure 8-1 are tall and short.

A tool called a *Punnett square* helps geneticists predict what kinds of offspring might result from a particular genetic cross. In Figure 8-1, a Punnett square shows the cross between the peas of the F1 generation. The alleles that each parent can contribute to the offspring are written along the sides of the square. All possible combinations of alleles that could result from the meeting of sperm and egg are drawn within the square.

If Mendel had used modern genetic notation and terminology, he might have analyzed his experiment to look like this (look to Figure 8-1 for reference if you need it):

1. The parental pea plants are purebred, so they have only one type of allele, but each individual plant has two alleles for each gene. The tall parent's alleles are shown as *TT*, and the

short parent's alleles are shown as *tt*. Because both of their alleles are the same, the parental pea plants are *homozygous* for the plant height trait (*homo-* means "same," and *zygous* comes from a Greek root that means "together").

2. Each parental pea plant gives one allele to each offspring. Because the parentals are purebred, they can give only one type of allele. Tall pea plant parents always give a copy of the tall allele *(T)* to offspring, and short parents always give a copy of the short allele *(t)*.

3. The sperm and egg (also called *gametes*) of the parents combine, giving their F1 offspring two alleles for the height gene. All the F1 offspring have one copy of each allele, so their alleles are written as *Tt*. Because their alleles are different, the F1 pea plants are *heterozygous* for the plant height trait (*hetero-* means "other"). Although the F1 plants are heterozygous, they should all look tall because the tall allele is dominant to the short allele. This is exactly what Mendel saw — the short trait from his parentals seemed to disappear in the F1 generation.

4. When F1 plants are crossed, they can each make two kinds of gametes — those that carry a dominant allele and those that carry a recessive allele. To figure out all the possible combinations of offspring the F1 plants could have, you use a Punnett square like the one shown in Figure 8-1. Write the two types of gametes each parent makes along the sides of the square. Figure out the possible genotypes of the offspring by filling in the squares with the different combinations of gametes.

5. The completed Punnett square in Figure 8-1 predicts that the F2 offspring will have three different genotypes: *TT, Tt,* and *tt*. For every one *TT* offspring, there should be two *Tt* offspring and one *tt* offspring. In other words, the *genotypic ratio* (the ratio of expected numbers of genotypes for the cross) predicted for the F2 generation is 1:2:1 for *TT:Tt:tt*.

6. The tall allele is dominant to the short allele, so F2 plants that are *TT* or *Tt* will be tall, and only F2 plants that are *tt* will be short. So, the Punnett square predicts that for every three tall plants, there'll be just one short plant. In other words, the *phenotypic ratio* (the ratio of expected numbers of phenotypes for the cross) for the F2 generation is 3:1 for tall:short. This is precisely what Mendel saw — for every one short plant he saw in his F2 generation, he saw about three tall ones.

Genetic Engineering

Gregor Mendel's pea plant experiments began a scientific exploration into the mysteries of heredity that continues to this day. After Mendel showed that traits were controlled by hereditary factors that pass from one generation to the next, scientists were determined to figure out the nature of these factors and how they were transmitted. They discovered the presence of DNA in cells, observed the movement of chromosomes during cell division, and conducted experiments demonstrating that DNA is in fact the hereditary material.

Almost 100 years after Mendel, James Watson and Francis Crick figured out that DNA was a double helix and proposed how it might be copied. Scientists deciphered the genetic code and explored how to work with it in the lab. During the last 40 years, scientists have developed an amazing array of tools to read DNA, copy it, cut it, sort it, and put it together in new combinations. The power of this DNA technology is so great that scientists have even determined the sequence of all the chromosomes in human cells as part of the Human Genome Project. A new world of human heredity is now open for exploration as scientists seek to understand the meanings hidden within human DNA — what they find out will likely change the way we see ourselves and our place in the world.

In this section, we get you acquainted with what is involved in DNA technology.

Understanding what's involved in DNA technology

For years the very structure of DNA made studying it rather challenging. After all, DNA is incredibly long and very tiny. Fortunately, the advent of *DNA technology*, the tools and techniques used for reading and manipulating the DNA code, has made working with DNA much easier. Scientists can even combine DNA from different organisms to artificially create materials such as human proteins or to give crop plants new characteristics. They can also compare different versions of the same gene to see exactly where disease-causing variations occur.

The following sections break down the various aspects of DNA technology so you can see how they all combine to provide a window into the very essence of existence.

Cutting DNA with restriction enzymes

Scientists use *restriction enzymes*, essentially little molecular scissors, in the lab to cut DNA into smaller pieces so they can analyze and manipulate it more easily. Each restriction enzyme recognizes and can attach to a certain sequence on DNA called a *restriction site*. The enzymes slide along the DNA, and wherever they find their restriction site, they cut the DNA helix.

Figure 8-2 shows how a restriction enzyme can make a cut in a circular piece of DNA and turn it into a linear piece.

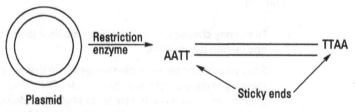

FIGURE 8-2: Restriction enzymes.

Combining DNA from different sources

After DNA has been chopped into smaller, more workable bits, scientists can combine pieces of DNA to change the characteristics of a cell. For example, they can put genes into crop plants to make them resistant to pesticides or to increase their nutritional value. This manipulation of a cell's genetic material in order to change its characteristics is called *genetic engineering*.

Because the DNA from all cells is essentially the same, scientists can even combine DNA from very different sources. For example, human DNA can be combined with bacterial DNA.

REMEMBER

When a DNA molecule contains DNA from more than one source, it's called *recombinant DNA*.

If a recombinant DNA molecule containing bacterial and human genes is put into bacterial cells, the bacteria read the human genes like their own and begin producing human proteins that scientists can use in medicine and scientific research. Table 8-1 lists a few useful proteins that are made through genetic engineering.

TABLE 8-1 Some Beneficial Genetically Engineered Proteins

Protein	Benefit
Alpha-interferon	Used to shrink tumors and treat hepatitis
Beta-interferon	Used to treat multiple sclerosis
Human insulin	Used to treat people with diabetes as a safer alternative to pig insulin
Tissue plasminogen activator (tPA)	Given to patients who've just had a heart attack or stroke to dissolve the blockage that caused the attack

Here's how scientists go about putting a human gene into a bacterial cell:

1. **First, they choose a restriction enzyme that forms sticky ends when it cuts DNA.**

   Sticky ends are pieces of single-stranded DNA that are complementary to other pieces of single-stranded DNA. Because they're complementary, the pieces of single-stranded DNA can stick to each other by forming hydrogen bonds. For example, the sticky ends shown in Figure 8-2 have the sequences 5'AATT3' and 3'TTAA5'. A and T are complementary base pairs, so these ends can form hydrogen bonds and stick to each other.

2. **Next, they cut the human DNA and bacterial DNA with the same restriction enzyme.**

   When you cut bacterial DNA and human DNA with the same restriction enzyme, all the DNA fragments have the same sticky ends.

3. **Then they combine human DNA and bacterial DNA.**

   Because the two types of DNA have the same sticky ends, some of the pieces stick together.

4. **Finally, they use the enzyme DNA ligase to seal the sugar-phosphate backbone between the bacterial and human DNA.**

   DNA ligase forms covalent bonds between the pieces of DNA, sealing together any pieces that are combined.

Genetically modified organisms

Genetically modified organisms (abbreviated as GMOs and sometimes called *genetically engineered organisms* or *transgenic organisms*) contain genes from other species that were introduced using recombinant DNA technology. GMOs are a hot topic these days due to the controversy surrounding genetically modified crop plants and farm animals. The sections that follow take a look at both sides of the "Are GMOs good or bad?" debate.

Why GMOs are beneficial

Genetic modification has its upsides. It not only makes growing crops easier but can also boost the profitability of those crops. And it may even help improve human health. Here are some specific scenarios that illustrate how GMOs can be beneficial:

» **If crop plants are given genes to resist herbicides and pesticides, a farmer can spray the fields with those chemicals, killing only the weeds and pests, not the crop plant.** This is much easier and less time-consuming than labor-intensive weeding. It can also increase crop yields and profits for the farmer.

» **If crop plants or farm animals raised for human consumption are given genes to improve their nutrition, people could be healthier.** Improved nutrition in crop plants could be a huge benefit in poor countries where malnutrition stunts the growth and development of children, making them more susceptible to disease. One of the most famous examples of improved nutrition through genetic engineering is the creation of "golden rice" — rice that has been engineered to make increased amounts of a nutrient that's necessary for vitamin A production. According to the World Health Organization, vitamin A deficiencies cause 250,000 to 500,000 children to go blind each year. The company that produced golden rice is giving the rice to poor countries for free so they can grow it for themselves and make it available to people who need it.

» **If farm animals raised for human consumption are given genes to increase their yield of meat, eggs, and milk, more food may be available for the growing human population, and these greater yields may also increase profits for farmers.** Currently, many dairy cows are given recombinant bovine growth hormone (rBGH) to increase their milk production. BGH is a normal growth hormone found in cows; rBGH is a slightly altered version that's produced by genetically engineered bacteria. When rBGH is given to cows, the animals' milk production increases by 10 to 15 percent.

Why GMOs cause concern

What make GMOs so controversial are the ethical concerns. The list of concerns surrounding genetic modification is long and so serious that some countries in the European Union have banned the sale of foods containing products from GMOs. The concerns expressed include the following:

» **The use of GMOs in agriculture unfairly benefits big agricultural companies and pushes out smaller farmers.** Companies that produce seeds for genetically engineered crops retain patents on their products. The prices on these seeds can be much higher than for traditional crops, giving large agricultural companies an advantage in the marketplace. This issue is particularly worrisome when large agricultural companies from rich nations start competing in the global economy with smaller farmers from poor countries.

» **The use of GMOs in agriculture encourages unsound environmental practices and discourages best farming practices.** Farmers who plant crops engineered for pesticide or herbicide resistance use chemicals rather than manual labor to control weeds and pests. Not only do these pesticides and herbicides affect the health of plants and animals living in the area around farms, but they can also get into the drinking water and possibly affect human health. Also, large-scale plantings of just a few species of plants decrease the genetic diversity in food species and put the food supply at risk for large-scale catastrophes should one of the crop species fail.

» **Animals that are engineered to produce more milk, eggs, or meat may be at greater risk for health problems.** Cows treated with rBGH to increase milk production get more infections in their milk ducts and have to be treated with antibiotics more often. Overuse of antibiotics is a human health concern because it reduces the effectiveness of antibiotics on bacteria that cause human infections.

» **Cross-pollination between genetically engineered plants and wild plants can spread resistant genes into wild plants.** Farmers can put up fences, but wind blows all over the place. If a crop plant that contains a gene for herbicide resistance can pollinate a wild plant, then the wild plant could pick up that gene, creating a weed species that can't be controlled.

>> **Increased levels of bovine hormones in dairy products may have effects on humans who drink the milk.** When rBGH is injected into cows to pump up their milk production, the levels of IGF-1 (an insulin-like protein) in their bodies and milk increase. Human bodies also make IGF-1, and increased levels of this hormone have been found in patients with some types of cancer. People are worried that increased IGF-1 in milk from hormone-treated cows may put them at greater risk for cancer, but no clear link has yet been found between IGF-1 in milk and human cancer.

>> **Genetic modification of foods may introduce allergens into foods, and labeling may not be sufficient to protect the consumer.** People who have food allergies have to be very careful about which foods they eat. However, if foods contain products from GMOs, it's possible that the introduced genes produced a product that's not indicated on the food label.

>> **Fear of "unnatural" practices and new technologies makes people afraid of GMOs and lowers their value in the marketplace.** Some people see humans as becoming out of balance with the rest of nature and think we need to slow down and try to leave less of a footprint on the world. For some, this belief includes rejecting technology that alters organisms from their natural state.

Reading a gene with DNA sequencing

REMEMBER

DNA sequencing, which determines the order of nucleotides in a DNA strand, allows scientists to read the genetic code so they can study the normal versions of genes. It also allows them to make comparisons between normal versions of a gene and disease-causing versions of a gene. After they know the order of nucleotides in both versions, they can identify which changes in the gene cause the disease.

As you can see later in the chapter in Figure 8-3, DNA sequencing uses a special kind of nucleotide, called ddNTP (short for dideoxyribonucleotide triphosphate). Regular DNA nucleotides and ddNTPs are somewhat similar, but the ddNTPs are different enough that they stop DNA replication. When a ddNTP is added to a growing chain of DNA, DNA polymerase can't add any more nucleotides. DNA sequencing uses this chain interruption to determine the order of nucleotides in a strand of DNA.

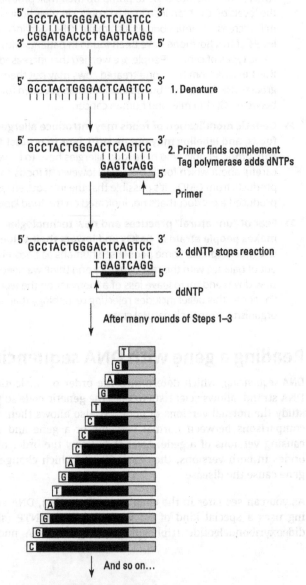

FIGURE 8-3: DNA sequencing.

Most DNA sequencing done today is *cycle sequencing,* a process that creates partial copies of a DNA sequence, all of which are stopped at different points. After the partial copies are made, scientists load them into a machine that uses a process called *gel electrophoresis.* After the partial copies of DNA are placed in a gel matrix, an electrical current is applied to the gel, causing the DNA pieces to separate out by size. The gel matrix used in DNA sequencing can separate pieces of DNA that differ by one nucleotide. As the partial sequences pass through the machine, a laser reads a fluorescent tag on each ddNTP, noting the DNA sequence.

Mapping the genes of humanity

The *Human Genome Project* (HGP) was a hugely ambitious task to determine the nucleotide sequence of all the DNA in a human cell. To give you an idea just how ambitious this project was, when it was first proposed in 1985, the pace of DNA sequencing was so slow that it would've taken 1,000 years to sequence the 23 unique human chromosomes. Fortunately, scientists cooperated and technology improved during the project, allowing the majority of the human genome to be sequenced by 2003. (A *genome* is the total collection of genes in a species.)

TIP

If you're wondering why the HGP is a big deal, think of it this way. If you were a researcher and you wanted to study a specific human gene, first you'd have to know what chromosome it "lived" on. The map of nucleotide sequences created by the HGP is a huge step forward in providing the "address" of each human gene. Armed with a road map of where every gene is located, researchers can turn their attention toward making good use of that information, like seeking out the genes that cause disease.

The HGP and the technological advances that came along with it resulted in many other current and potential benefits to society, including

>> Drugs designed to best treat an individual person with minimal side effects

>> Earlier detection of disease

>> Exploration of microbial genomes for identification of species that can be used to produce new biofuels or clean up pollution

>> Comparison of DNA from crime scenes to that of suspects in order to help determine likely guilt or innocence

>> Study of the evolutionary relationships of life on Earth

Chapter **9**

Biological Evolution

f you've ever been to a museum, you've probably seen fossilized bones or tools from ancient ancestors. These objects are evidence of how humans have changed and expanded their knowledge over the millenia. In other words, they provide perspective on how the human species has evolved. But what was the starting point of evolution, and from what did the earliest humans evolve?

This chapter tells you about the beliefs people once had regarding evolution; how Charles Darwin came up with his theory of biological evolution; and what the current thoughts are on the origin of species, how humans have evolved, and how life on Earth began.

What People Used to Believe

From the time when ancient Greece was the world's cultural hot spot until the early 1800s, philosophers, scientists, and the general public believed that plants and animals were specially created at one time and that new species hadn't been introduced since then. (You could call this way of thinking *fundamentalism.*) In this view, every living thing was created in its ideal form by the hand of God for a special purpose. Aristotle classified all living things into a "great chain of being" from simple to complex, placing human beings at the top of the chain, just under the angels and very close to God.

People also thought the Earth and the universe it occupied were unchanging, or *static*, throughout time. They believed that God created the Earth, the stars, and the other planets all at once and that nothing had changed since the dawn of creation. These ideas held through much of human history.

Beginning in the 15th century and continuing into the 18th century, explorers, scientists, and naturalists made new discoveries that challenged the old ideas of a static universe:

>> Various explorers fell upon the New World (the Western Hemisphere of the Earth). The New World revealed many different species of living things, including new races of people, that were previously unknown. These puzzles raised questions about a literal interpretation of the creation story in the *Bible*'s book of Genesis.

>> William Smith, a British surveyor, discovered that the ground consisted of layers of different material and that different types of fossils could be found in each layer. The deeper he went into the layers, the more different the fossils appeared from the plants and animals that lived in Britain at the time.

>> Georges Cuvier, a French anatomist, demonstrated that fossil bones found in Europe, such as those of wooly mammoths, could be recognized as very similar to existing species, such as elephants, but were clearly not from anything currently living.

>> James Hutton, a Scottish geologist, proposed that the Earth was very ancient and that its surface was constantly changing due to erosion, the depositing of sediment, the uplift of mountains, and flooding. His idea, called *uniformitarianism,* was that the processes he observed on the Earth in the 1700s were the same processes that had occurred on the Earth since its creation.

Charles Darwin: Challenging Age-Old Beliefs

Charles Darwin was a gentleman from the English countryside who set out on a seafaring journey on the HMS *Beagle* in 1831 as the ship's naturalist. His observations led to the creation of two of the most important biological theories of all time: biological evolution and natural selection.

Owing it all to the birds

While traveling on the HMS *Beagle*, Darwin visited the Galapagos Islands, which lie nearly 600 miles off the western coast of South America. He was amazed to find a variety of species that were similar to those in South America yet different in ways that seemed to make them especially suited to the unique environment of the isolated islands.

REMEMBER

Characteristics of organisms that make them suited to their environment are called *adaptations*.

Darwin chose to focus his attention on the Galapagos Islands' finches (a type of bird). Each island had its own unique species of finch that was distinct from the other species and from the finches on the mainland. In South America, finches ate only seeds. On the islands, some finches ate seeds, others ate insects, and some even ate cactuses. The beak of each type of finch seemed exactly suited to its food source.

Darwin thought that all the finches had a common ancestor from mainland South America that either flew or floated to the newly formed islands, perhaps during occasional storms. The islands are far enough apart from each other that finches can't really travel between them, so the different populations are geographically isolated from each other. *Geographic isolation* means they also can't mate with each other and combine their genes.

Darwin proposed that each type of island had unique conditions and that these unique conditions favored certain traits over others. Birds whose traits made them more successful at obtaining food were more likely to survive and reproduce, passing their genes and traits on to their offspring. Over time, the characteristics of the island birds shifted away from those of their mainland ancestor toward characteristics that better suited their new home. Eventually, the island birds became so different from their ancestors, and from each other, that they were unique species.

Darwin's theory of biological evolution

Biological evolution refers to the change of living things over time. (It's not the same as *evolution*, which simply means change.) Darwin introduced the world to this concept in his 1854 work, *On the Origin of Species*. In this book, Darwin proposed that living things descend from their ancestors but that they can change over time. In other words, Darwin believed in *descent with modification*.

REMEMBER

As changes occur in living things, species that don't adapt to changing environmental conditions may become *extinct*, or disappear. Species that accumulate enough changes may become so different from related organisms that they become a new species because they can no longer successfully mate with related populations; this process is referred to as *speciation*.

The idea of natural selection

Darwin concluded that biological evolution occurred as a result of *natural selection*, which is the theory that in any given generation, some individuals are more likely to survive and reproduce than others. When a particular trait improves the survivability of an organism, the environment is said to favor that trait or naturally select for it. Natural selection therefore acts against unfavorable traits.

TIP

The theory of natural selection is often referred to as "survival of the fittest." Biological fitness is basically your ability to produce offspring. So, survival of the fittest really refers to the passing on of those traits that enable individuals to survive and successfully reproduce.

Natural selection versus artificial selection

Darwin compared his theory of natural selection with the artificial selection that results from selective breeding in agriculture.

>> *Artificial selection* occurs when people choose plants or animals and breed them for certain desired characteristics.

>> *Natural selection* occurs when environmental factors "choose" which plants or animals will survive and reproduce. If a visual predator, such as an eagle, is cruising for its lunch, the individuals that it can see most easily are likely to be eaten. If the eagle's prey is mice, which can be white or dark colored (see Figure 9-1a), and the mice live in the forest against dark-colored soil, then the eagle is going to be able to see the white mice more easily. Over time, if the eagles in the area keep eating more white mice than dark mice (see Figure 9-1b), then more dark mice are going to reproduce. Dark mice have genes that specify dark-colored fur, so their offspring will also have dark fur. If the eagle continues to prey upon mice in the area, the population of mice in the forest will gradually begin to have more dark-colored individuals than white individuals (see Figure 9-1c).

In this example, the eagle is the *selection pressure*: an environmental factor that causes some organisms to survive (the dark-colored mice) and others not to survive (the white-colored mice). A selection pressure gets its name from putting "pressure" or stress on the individuals of the population.

Organisms with the best-suited characteristics for their environment are more likely to survive and reproduce. This is the heart of natural selection.

REMEMBER

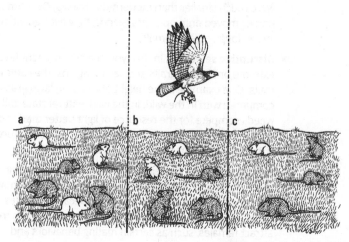

Illustration by Kathryn Born, MA

FIGURE 9-1: Natural selection in action.

Four types of natural selection

Natural selection can cause several different types of changes in a population. How the population changes depends upon the particular selection pressure the population is under and which traits are favored in that circumstance. Individuals within a population may evolve to be more similar to or more different from each other depending on the specific circumstances and selection pressures.

The four types of natural selection are as follows:

REMEMBER

>> **Stabilizing selection:** This type eliminates extreme or unusual traits. Individuals with the most common traits are considered best adapted, which maintains the frequency of common traits in the population. The size of human babies, for example, remains within a certain range due to stabilizing selection.

>> **Directional selection:** In this type, traits at one end of a spectrum of traits are selected for, whereas traits at the other end of the spectrum are selected against. Over generations, the selected traits become common, and the other traits become more and more extreme until they're eventually phased out. The biological evolution of horses is a good example of directional selection. Ancestral horse species were built for moving through wooded areas and were much smaller than modern-day horses. Over time, as horses moved onto open grasslands, they evolved into much larger, long-legged animals.

>> **Disruptive selection:** In this type, the environment favors extreme or unusual traits and selects against the common traits. One example is the height of weeds in lawn grass compared with in the wild. In the wild, natural state, tall weeds compete for the resource of light better than short weeds. But in lawns, weeds have a better chance of surviving if they remain short because grass is kept short.

>> **Sexual selection:** Females increase the fitness of their offspring by choosing males with superior fitness; females are therefore concerned with quality. Because females choose their mates, males have also developed traits to attract females, such as certain mating behaviors and bright coloring.

REMEMBER

Biological evolution happens to populations, not individuals. Individuals live or die and reproduce or don't reproduce depending on their circumstances. But individuals themselves can't evolve in response to a selection pressure. Imagine a giraffe whose neck isn't quite long enough to reach the tastiest leaves at the top of the tree. That individual giraffe can't suddenly grow its neck longer to reach the leaves. However, if another giraffe in the herd has a longer neck, gets more leaves, grows better, and makes more calves that inherit his long neck, then future generations of giraffes in that area may have longer necks.

Evidence of Biological Evolution

Since Darwin first proposed his ideas about biological evolution and natural selection, many different lines of research from many different branches of science have produced evidence supporting his belief that biological evolution occurs in part due to natural selection.

REMEMBER

Because a great amount of data supports the idea of biological evolution through natural selection, and because no scientific evidence has yet been found to prove this idea false, this idea is considered a scientific theory.

Biochemistry

The fundamental *biochemistry* (the basic chemistry and processes that occur in cells) of all living things on Earth is incredibly similar, showing that all Earth's organisms share a common ancestry.

Case in point: All living things store their genetic material in DNA and build proteins out of the same 20 amino acids. Regardless of whether the organisms are flowers taking in carbon dioxide from the air, water from the soil, and light from the sun; lions chomping down a wildebeest; or humans consuming a gourmet meal cooked by Wolfgang Puck himself, all organisms convert food sources to energy and store that energy in ATP. That stored energy is then used to power cellular processes such as the production of proteins, which is directed by the genes on strands of DNA.

Comparative anatomy

Comparative anatomy — which looks at the structures of different living things to determine relationships — has revealed that the various species on Earth evolved from common ancestors.

As you can see in Figure 9-2, the skeletons of humans, cats, whales, and bats, for example, are amazingly similar even though these animals live unique lifestyles in very different environments. From the outside, the arm of a human, the front leg of a cat, the flipper of a whale, and the wing of a bat seem very different, but when you look at the bones within them, you see that they all contain the same ones — an upper "arm," an elbow, a lower "arm," and five "fingers." The only differences in these bones are their size and shape. Scientists call similar structures such as these *homologous structures* (*homo*- means "same"). The best explanation for these homologous structures is that all four mammals are descended from the same ancestor — an idea that's supported by the fossil record.

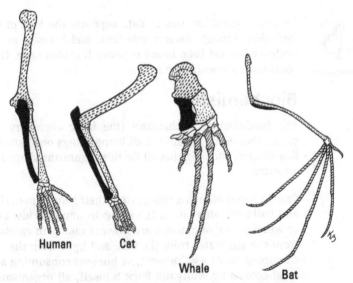

Human   Cat

Whale

Bat

Illustration by Kathryn Born, MA

FIGURE 9-2: Comparative anatomy of the bones in front limbs of humans, cats, whales, and bats.

Geographic distribution of species

REMEMBER

How populations of species are distributed around the globe helps solidify Darwin's theory of biological evolution. In fact, the science of *biogeography*, the study of living things around the globe, allows scientists to make testable predictions about biological evolution. Basically, if biological evolution is real, you'd expect groups of organisms that are related to each other to be clustered near each other because related organisms come from the same common ancestor. (An exception to this prediction is that migratory animals could travel far from their relatives.) On the other hand, if biological evolution isn't real, there's no reason for related groups of organisms to be found near each other. For example, a creator could scatter organisms randomly all over the planet, or groups of organisms could arise independently of other groups in whatever environments suited them best.

When biogeographers compare the distribution of organisms living today, they find that species are distributed around the Earth in a pattern that reflects their genetic relationships to one another.

When Darwin compared the finches on the Galapagos Islands with those on mainland South America, the unique types of finches on the Galapagos Islands led him to hypothesize that the islands had been colonized by finches from the mainland. This hypothesis was later supported when modern scientists performed a genetic analysis of the Galapagos Islands' finches and were able to demonstrate their relationship to each other and to their mainland ancestors.

Since Darwin's time, many other examples have been found that illustrate how geographic distribution has influenced the biological evolution of organisms.

Molecular biology

Molecular biology is the branch of biology that focuses on the structure and function of the molecules that make up cells. With it, biochemists have been able to compare the structures of proteins from many different species and use the similarities to create *phylogenetic trees* (essentially family trees) that show the proposed relationships between organisms based on similarities between their proteins.

With the development of DNA technology that allows for reading of the actual gene sequence in DNA, modern scientists have also been able to compare gene sequences among species. Some proteins and gene sequences are similar between very distantly related organisms, indicating that they haven't changed in millions of years.

Fossil record

The *fossil record* (all the fossils ever found and the information gained from them) shows detailed evidence of the changes in living things over time. During Darwin's day, the science of *paleontology*, which studies prehistoric life through fossil evidence, was just being born. Since Darwin's time, paleontologists have been busy filling in gaps left in the fossil record in order to explain the evolutionary history of organisms.

Hundreds of thousands of fossils have been found, showing the changing forms of organisms. For some types of living things, such as fish, amphibians, reptiles, and primates, the fossil record depiction of the changes from one form of the organism to another is so complete that it's hard to say where one species ends and the next one begins.

REMEMBER

Based on the fossil record, paleontologists have established a solid timeline of the appearance of different types of living things, beginning with the appearance of prokaryotic cells and continuing through modern humans.

Observable data

Biological evolution can be measured by studying the results of scientific experiments that measure evolutionary changes in the populations of organisms that are alive today. In fact, you need only look in the newspaper or hop online to see evidence of biological evolution in action in the form of antibiotic-resistant bacteria.

In the 1940s, when people first started using antibiotics to treat infections, most strains of the bacterium *Staphylococcus aureus* (*S. aureus*) could be killed by penicillin. By using antibiotics, people applied a strong selection pressure to the populations of the *S. aureus* bacteria. The fittest *S. aureus* bacteria were those that could best withstand the penicillin. The bacteria that couldn't withstand the penicillin died, and the resistant bacteria multiplied. Today, most populations of *S. aureus* are resistant to natural penicillin.

Radioisotope dating

Radioisotope dating indicates that the Earth is 4.5 billion years old — that's plenty old enough to allow for the many changes in Earth's species due to biological evolution. *Isotopes* are different forms of the atoms that make up matter on Earth (see Chapter 2 for more on isotopes). Some isotopes, called *radioactive isotopes,* discard particles over time and change into other elements. Scientists know the rate at which this radioactive decay occurs, so they can take rocks and analyze the elements within them. Using the known rates of radioactive decay and the types of elements that were originally present in the rocks, scientists can calculate how long the elements in a particular rock have been discarding particles — in other words, they can figure out the age of the rock (including rocks with fossils).

Evolution versus Creationism

Virtually all scientists today agree that biological evolution happens and that it explains many important observations about living things, but many nonscientists don't believe in biological evolution and are often violently opposed to it. They prefer to take the *Bible*'s creation story literally. These wildly differing viewpoints have led to one of the great debates of all time: Which is correct, evolution or creationism?

REMEMBER

The idea of biological evolution has inspired so much controversy over the years in large part because many people think it contradicts the Christian view of humanity's place in God's design.

At the root of the controversy about biological evolution seems to be this question: If living things developed in all their wonderful complexity due to natural processes and without the direct involvement of God, what does that do to man's place in the world? Is mankind not "special" to God?

But are biological evolution and religious faith necessarily in conflict? Many religious figures and scientists don't think so. In fact, many scientists have strong religious beliefs, and many religious leaders have come forward to say that they believe in biological evolution.

Ultimately, each person's beliefs are under his or her own control. But scientists stress the difference between beliefs, or faith, and science.

>> Science is an attempt to explain the natural world based on observations made with the five senses. Scientific ideas, or hypotheses, must be testable — able to be proven false — by observation and experimentation.

>> The existence of God isn't within the scientific realm. God is widely believed to be a supernatural being, outside the workings of the natural world. Belief in the existence of God is therefore a matter of faith.

» **Creating vaccinations, antibiotics, and treatments for genetic defects**

Chapter **10**
Ten Great Biology Discoveries

Get ready to dive into ten of the most important biology discoveries to date. We list them in no particular order because they've all made a significant impact on the advancement of biology as a science and increased what people know and understand about the living world.

Seeing the Unseen

Before 1675, people believed the only living things that existed were the ones they could see. That year, a Dutch cloth merchant named Antony van Leeuwenhoek discovered the microbial world by peering through a homemade microscope. Van Leeuwenhoek was the first person to see bacteria, which he described as little animals that moved about here, there, and everywhere. His discovery of a previously unseen universe not only turned people's worldviews inside out but also laid the foundation for the understanding that microbes cause disease.

Creating the First Antibiotic

People had very few tools to combat bacterial infections until Alexander Fleming discovered the antibacterial properties of penicillin in 1928. Fleming was studying a strain of Staphylococcus bacteria when some of his petri dishes became contaminated with Penicillium mold. To Fleming's surprise, wherever the Penicillium grew on the petri dish, the mold inhibited the growth of the Staphylococcus bacteria.

The compound penicillin was purified from the mold and first used to treat infections in soldiers during World War II. Soon after the war, the "miracle drug" was used to treat infections in the general public, and the race to discover additional antibiotics was on.

Protecting People from Smallpox

Would you believe that the idea of vaccinating people against diseases such as smallpox, measles, and mumps originated in ancient China? Healers there ground up scabs taken from a smallpox survivor into a powder and blew this dust into the nostrils of their patients. Gross as this may sound, these ancient healers were actually inoculating their patients to help prevent the spread of the disease.

Defining DNA Structure

In 1953 James Watson and Francis Crick figured out how a code could be captured in the structure of DNA molecules, opening the door to an understanding of how DNA carries the blueprints for proteins. They proposed that DNA is made of two nucleotide chains running in opposite directions and held together by hydrogen bonds between the nitrogenous bases. Using metal plates to represent the bases, they built a giant model of DNA that was accepted as correct almost immediately.

Finding and Fighting Defective Genes

On August 24, 1989, scientists announced their discovery of the first known cause of a genetic disease: They found a tiny deletion from a gene on Chromosome 7 that resulted in the deadly genetic disease cystic fibrosis. This identification of a genetic defect, and the realization that this defect causes a disease, opened the floodgates of genetic research. Since that fateful day, the genes for other diseases, such as Huntington's disease, inherited forms of breast cancer, sickle cell anemia, Down syndrome, Tay-Sachs disease, hemophilia, and muscular dystrophy, have been found. Genetic tests for these diseases are available to detect whether an unborn baby has a defective gene or whether two potential parents would likely produce an affected baby. And knowing what causes the diseases enables researchers to focus on ways to possibly cure them.

Discovering Modern Genetic Principles

Gregor Mendel, a mid-19th century Austrian monk, used pea plants to perform the fundamental studies of heredity that serve as the basis for genetic concepts to this day. Because pea plants have a number of readily observable traits — smooth peas versus wrinkled peas, tall plants versus short plants, and so on — Mendel was able to observe the results of cross-pollinating and growing various varieties of pea plants.

Through his experiments, Mendel established that genetic factors are passed from parents to offspring and remain unchanged in the offspring so that they can be passed on again to the next generation. Although his work was done before the discovery of DNA and chromosomes, the genetic principles of dominance, segregation, and independent assortment that Mendel originally defined are still used to this day.

Evolving the Theory of Natural Selection

Charles Darwin's study of giant tortoises and finches on the Galapagos Islands led to his famous theory of natural selection (also known as "survival of the fittest"), which he published in his 1859 book titled *On the Origin of Species*. The main point of Darwin's theory is that organisms with traits that are better suited to the conditions in which they live are more likely to survive and reproduce, passing on their traits to future generations. These better-suited variations tend to thrive in the given area, whereas less-suited variations of the same species either don't do as well or just die off. Thus, over time, the traits seen in a population of organisms in a given area can change. The significance of Darwin's theory of natural selection can be seen today in the evolution of antibiotic-resistant strains of bacteria.

Formulating Cell Theory

In 1839, zoologist Theodor Schwann and botanist Matthias Schleiden were talking at a dinner party about their research. As Schleiden described the plant cells he'd been studying, Schwann was struck by their similarity to animal cells. The similarity between the two types of cells led to the formation of *cell theory*, which consists of three main ideas:

>> All living things are made of cells.

>> The cell is the smallest unit of living things.

>> All cells come from preexisting cells.

Moving Energy through the Krebs Cycle

The *Krebs cycle*, named for German-born British biochemist Sir Hans Adolf Krebs, is the major metabolic process that occurs in all living organisms. This process results in the transfer of energy to adenosine triphosphate (ATP), which all living things use to fuel their cellular functions. Defining how organisms use energy at the cellular level opened the door for further research on metabolic disorders and diseases.

Amplifying DNA with PCR

In 1983, U.S. chemist Kary Mullis discovered the *polymerase chain reaction* (PCR), a process that allows scientists to make numerous copies of DNA molecules that they can then study. Today, PCR is used for:

» Making lots of DNA for sequencing

» Finding and analyzing DNA from very small samples for use in forensics

» Detecting the presence of disease-causing microbes in human samples

» Producing numerous copies of genes for genetic engineering

Index

F

F1 generation, 131
F2 generation, 131
family, 12. *See also* genetics
fatty acids, 59
feedback inhibition, 56
fertilization, 91
First Law of Thermodynamics, 58, 59, 122
fission, 93
5′ cap, 101–102
flagella, 49
Fleming, Alexander (scientist), 158
fluid-mosaic model, 45–46
food. *See also* cellular respiration
  breaking down with cellular respiration, 67–68
  consuming for matter and energy, 60–61
  finding compared with producing, 61–62
food chain, 121
forest biomes, 113
fossil record, 154
fragmentation, 94
frameshift mutations, 108
freshwater biomes, 113
fundamentalism, 145–146
fungi, 10

G

G_1 phase, 80
G_2 phase, 80–81
gametes, 79, 85
gel electrophoresis, 143
gene regulation, 109–110
genes
  about, 34–36, 133
  defined, 96, 131–132

discovery of genetic disease, 159
  mapping, 143
  reading with DNA sequencing, 141–143
genetic cross, 133–135
genetic disease, 159
genetic engineering
  about, 136
  DNA sequencing, 141–143
  DNA technology, 136–138
  genetically modified organisms (GMOs), 139–141
  Human Genome Project (HGP), 141–142
genetic information, of living things, 11
genetic principles, discovering, 159
genetic tests, 159
genetic variation, 90–93
genetically engineered organisms, 139–141
genetically modified organisms (GMOs), 139–141
genetics
  genetic cross, 133–135
  genetic engineering, 136–143
  heritable traits, 129–130
  key terms, 132–135
  Mendel's Laws of Inheritance, 130–132
genome, 143
genotype, 134
genus, 13
geographic distribution of species, 153
glucose, 31–32, 63
glycerol, 59
glycogen, 31
glycolysis, 66, 68
GMOs (genetically modified organisms), 139–14
Golgi apparatus, 43, 50–51
gonads, 79
grassland biomes, 113

About the Authors

Rene Fester Kratz, PhD, teaches cellular biology and microbiology. She is a member of the North Cascades and Olympic Science Partnership, where she helped create inquiry-based science courses for future teachers. Kratz is also the author of *Molecular and Cell Biology For Dummies* (Wiley) and *Microbiology The Easy Way.*

Donna Rae Siegfried has written about pharmaceutical and medical topics for 15 years in publications including *Prevention, Runner's World, Men's Health,* and *Organic Gardening.* She has taught anatomy and physiology at the college level. She is also the author of *Anatomy & Physiology For Dummies* (Wiley).

Publisher's Acknowledgments

Project Editor: Joan Friedman

Acquisitions Editor:
Lindsay Sandman Lefevere

Assistant Editor: David Lutton

Technical Editors:
Jeffrey S. Robertson,
Allison Thomas

Senior Editorial Manager:
Jennifer Ehrlich

Editorial Supervisor and Reprint Editor: Carmen Krikorian

Editorial Assistant:
Rachelle S. Amick

Senior Project Coordinator:
Kristie Rees

Production Editor: Siddique Shaik

Cover Photos: © ktsdesign/
Shutterstock

Take dummies with you everywhere you go!

Whether you are excited about e-books, want more from the web, must have your mobile apps, or are swept up in social media, dummies makes everything easier.

Find us online!

dummies.com

dummies

A Wiley Brand

Leverage the power

Dummies is the global leader in the reference category and one of the most trusted and highly regarded brands in the world. No longer just focused on books, customers now have access to the dummies content they need in the format they want. Together we'll craft a solution that engages your customers, stands out from the competition, and helps you meet your goals.

Advertising & Sponsorships

Connect with an engaged audience on a powerful multimedia site, and position your message alongside expert how-to content. Dummies.com is a one-stop shop for free, online information and know-how curated by a team of experts.

- Targeted ads
- Video
- Email Marketing
- Microsites
- Sweepstakes sponsorship

20 **MILLION** PAGE VIEWS
EVERY SINGLE MONTH

15 MILLION **UNIQUE**
VISITORS PER MONTH

43% OF ALL VISITORS ACCESS THE SITE
VIA THEIR MOBILE DEVICES

700,000 NEWSLETTER SUBSCRIPTION
TO THE INBOXES OF
300,000 UNIQUE INDIVIDUALS EVERY WEEK

# of dummies

Custom Publishing

Reach a global audience in any language by creating a solution that will differentiate you from competitors, amplify your message, and encourage customers to make a buying decision.

- Apps
- Books
- eBooks
- Video
- Audio
- Webinars

Brand Licensing & Content

Leverage the strength of the world's most popular reference brand to reach new audiences and channels of distribution.

For more information, visit **dummies.com/biz**

dummies
A Wiley Brand

PERSONAL ENRICHMENT

Staying Sharp
9781119187790
USA $26.00
CAN $31.99
UK £19.99

Facebook
9781119179030
USA $21.99
CAN $25.99
UK £16.99

Guitar
9781119293354
USA $24.99
CAN $29.99
UK £17.99

Investing
9781119293347
USA $22.99
CAN $27.99
UK £16.99

Beekeeping
9781119310068
USA $22.99
CAN $27.99
UK £16.99

Digital Photography
9781119235606
USA $24.99
CAN $29.99
UK £17.99

Meditation
9781119251163
USA $24.99
CAN $29.99
UK £17.99

Pregnancy
9781119235491
USA $26.99
CAN $31.99
UK £19.99

Samsung Galaxy S7
9781119279952
USA $24.99
CAN $29.99
UK £17.99

iPhone
9781119283133
USA $24.99
CAN $29.99
UK £17.99

Crocheting
9781119287117
USA $24.99
CAN $29.99
UK £16.99

Nutrition
9781119130246
USA $22.99
CAN $27.99
UK £16.99

PROFESSIONAL DEVELOPMENT

Windows 10
9781119311041
USA $24.99
CAN $29.99
UK £17.99

AutoCAD
9781119255796
USA $39.99
CAN $47.99
UK £27.99

Excel 2016
9781119293439
USA $26.99
CAN $31.99
UK £19.99

QuickBooks 2017
9781119281467
USA $26.99
CAN $31.99
UK £19.99

macOS Sierra
9781119280651
USA $29.99
CAN $35.99
UK £21.99

LinkedIn
9781119251132
USA $24.99
CAN $29.99
UK £17.99

Windows 10
9781119310563
USA $34.00
CAN $41.99
UK £24.99

SharePoint 2016
9781119181705
USA $29.99
CAN $35.99
UK £21.99

Fundamental Analysis
9781119263593
USA $26.99
CAN $31.99
UK £19.99

Networking
9781119257769
USA $29.99
CAN $35.99
UK £21.99

Office 2016
9781119293477
USA $26.99
CAN $31.99
UK £19.99

Office 365
9781119265313
USA $24.99
CAN $29.99
UK £17.99

Salesforce.com
9781119239314
USA $29.99
CAN $35.99
UK £21.99

Coding
9781119293323
USA $29.99
CAN $35.99
UK £21.99

dummies.com

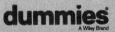
A Wiley Brand

Learning Made Easy

ACADEMIC

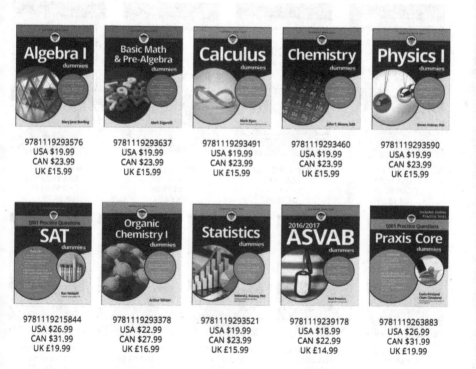

Algebra I
9781119293576
USA $19.99
CAN $23.99
UK £15.99

Basic Math & Pre-Algebra
9781119293637
USA $19.99
CAN $23.99
UK £15.99

Calculus
9781119293491
USA $19.99
CAN $23.99
UK £15.99

Chemistry
9781119293460
USA $19.99
CAN $23.99
UK £15.99

Physics I
9781119293590
USA $19.99
CAN $23.99
UK £15.99

SAT
9781119215844
USA $26.99
CAN $31.99
UK £19.99

Organic Chemistry I
9781119293378
USA $22.99
CAN $27.99
UK £16.99

Statistics
9781119293521
USA $19.99
CAN $23.99
UK £15.99

2016/2017 ASVAB
9781119239178
USA $18.99
CAN $22.99
UK £14.99

Praxis Core
9781119263883
USA $26.99
CAN $31.99
UK £19.99

Available Everywhere Books Are Sold

dummies.com

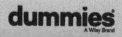
dummies
A Wiley Brand

Small books for big imaginations

9781119177173
USA $9.99
CAN $9.99
UK £8.99

9781119177272
USA $9.99
CAN $9.99
UK £8.99

9781119177241
USA $9.99
CAN $9.99
UK £8.99

9781119177210
USA $9.99
CAN $9.99
UK £8.99

9781119262657
USA $9.99
CAN $9.99
UK £6.99

9781119291336
USA $9.99
CAN $9.99
UK £6.99

9781119233527
USA $9.99
CAN $9.99
UK £6.99

9781119291220
USA $9.99
CAN $9.99
UK £6.99

9781119177302
USA $9.99
CAN $9.99
UK £8.99

Unleash Their Creativity

dummies.com